ADVANCED METHODS
FOR
SHEET METAL WORK

ADVANCED METHODS
for
SHEET
METAL
WORK

By William Cookson, C.Eng., F.I.Prod.E.

Managing Director:
Cookson Sheet Metal Developments Ltd.
and
C.S.M.D. International Ltd.,
Southampton

Sixth Edition
Revised, Enlarged and Metricated

OXFORD
TECHNICAL PRESS

First Edition	–	–	1941
Second Edition	–	–	1943
Third Edition	–	–	1944
Fourth Edition	–	–	1954
Fifth Edition	–	–	1964
Sixth (Metric) Edition	–		1975

ISBN 0 291 39427 2

MADE AND PRINTED BY OFFSET IN GREAT BRITAIN
BY WILLIAM CLOWES & SONS, LIMITED, LONDON, BECCLES AND COLCHESTER

CONTENTS

PREFACE TO THE SIXTH EDITION

The main object of this Sixth Edition of a book first published to meet the pressing needs of the war-time armaments factories in 1941 has been to convert every measurement and every calculation given in earlier editions into S.I. metric units. This has necessarily involved large-scale resetting in the second half of the book, and the opportunity has been taken to replace with more modern illustrations many of the older drawings in the chapters in question.

A second feature of this Sixth Edition is a new chapter describing an improved technique of cold roll-forming sheet metal of light gauge which has been developed over the past few years under the Author's direction in the firm which bears his name. The feature of this new technique is that the strip coming off the uncoiler is not held flat in the last two or three metres of its passage towards the first roll-stand on the line, but is allowed instead to fold or " drape " downwards towards the shaft of the first set of rolls. The advantage gained is that the metal, in thus assuming a natural free-flowing form as it approaches the first pair of rolls, is relieved of much of the tearing and scurfing stresses which have proved a problem in the conventional cold roll-forming process. Other advantages of the new technique are also discussed—supported as they now are by some ten years' practical experience gained in operational conditions in various parts of the world.

For the rest, the purpose of the book remains what it always was—to modernise and simplify the procedure of geometrical development for the engineer and draughtsman working with sheet-metal of light gauge, and to explain the use of methods of calculation in the laying-out of complex patterns which enable the dimensional accuracy of the finished product to be achieved more exactly than has hitherto been thought possible. Several of the mathematical formulæ evolved by the Author have been specified by the British Standards Institution (in BS 1549, Part 1) for use in pattern development ; while earlier editions of the book have been recommended for the use of students by the City and Guilds of London Institute.

WILLIAM COOKSON

ADVANCED METHODS FOR SHEET METAL WORK

Chapter One

METHODS OF PATTERN DEVELOPMENT

THE sheet metal worker who aspires to an all-round knowledge of his craft must of necessity be skilled in the art of drafting patterns from working drawings. It is not sufficient to have a superficial idea of the different methods of pattern cutting and then rely on " rule-of-thumb " methods when a difficult job has to be made. Draughtsmen are too often blamed for what are termed fantastic designs, simply because the average craftsman cannot mark-out the correct patterns for the work.

Simplifications of the designs are then often sought, and the desired modifications are made to the drawings to bring the job within the scope of the workshop. It must be realised, however, that the main duties of the draughtsmen are to give the dimensions and shape of the finished job. They are very little concerned with the surface development of the various parts, unless, perhaps, in a minor way when estimating the approximate sizes of the metal required in making out a material schedule. Particularly in the case of air-duct work, such as transformers and junction pieces, the complaint is often made in the shops that insufficient information is given on the drawings, and that joints, for example, should be more clearly shown by the use of auxiliary projections or similar views.

This means, in effect, that the given orthographic plans and elevations have to be drawn and the required additional views marked-out on the sheet metal by the craftsman. As a large amount of projection work is thus necessary, much valuable time is used. Confusion can also arise through the multiplicity of lines required in the process. Knowledge of pattern development, therefore, to be practical, must be of such a character that jobs can be marked-out as far as is possible directly from the given views on the blue print. To this end, any methods of drafting patterns which involve high labour costs must be discarded and new methods adopted.

Several new and simple methods are described in this work, and by their use the sheet metal worker should be able to tackle difficult drawings sent out from the drawing office and in this way help to raise the technical status of his craft. Even in these days of mass production the skilled template maker and others who can draft patterns correctly are in great demand, for much has yet to be done to make the art of pattern drafting as well known as it should be in the trade.

Too little attention, also, is paid to the necessary adjustments to be made to the pattern for the thickness of metal used, as this must be taken very much into account for accurate light engineering work. It is useless to develop a pattern if, when it is made up, the finished job does not conform to the exact size and shape as specified on the working drawing. In aircraft work, for instance, very fine limits must be adhered to, as these limits are insisted upon through rigid systems of inspection.

The days of the " hammer and chisel " are long past, and a good knowledge of mathematics is as necessary to the modern craftsman as to the people who draw the plans. A knowledge of the geometrical development of surfaces calls for some grounding in the principles of the different methods of drafting patterns in general use. There are three methods, of which the first is the development by parallel lines of patterns for articles shaped in the form of prisms and cylinders. The second is the radial-line method, used for obtaining the patterns for conical objects, and the third is the method of triangulation.

Of these three general methods the parallel-line and triangulation methods are the most widely used. The radial-line method is not applied as much as formerly, because, with the exception of the right cone and its frustums, all work of a conical nature is best tackled by the triangulation method.

In this work only a brief description and a few typical examples of parallel-line and radial-line development are given, as these present nothing new and are more fully dealt with elsewhere. The triangulation system described, however, is different from the methods hitherto in use in the trade. As the great merit of triangulation is its adaptability to the development of the majority of patterns, its use cannot be too highly stressed as a means of tackling most problems which may confront the craftsman in his daily tasks.

PARALLEL-LINE DEVELOPMENT

This method can be applied to the development of the patterns for elbows and tee-pieces for air-duct and similar work. The underlying principle of the system can be seen by studying the illustration shown in Fig. 1. This depicts the perspective view of a round pipe, with one end cut on the slant, placed on a horizontal plane. Drawn on its surface are a series of equally spaced parallel lines. An elevation of the pipe is shown projected on to an adjacent vertical plane ; the edge lines of the elevation are lettered and numbered A-0 and G-6 respectively.

The parallel lines on the elevation surface are the same length as those on the pipe, but they are not equally spaced. If the job were cut at the seam 0-A and the metal " un-rolled," we should obtain the pattern as it would appear on the flat sheet, as shown in Fig. 2 (*b*). Thus, if we first draw an elevation of the pipe as in Fig. 2, and draw parallel lines on its surface similar to those on the elevation in the perspective view, we can set-out a pattern by reproducing the parallel lines in their correct relation to each other and cutting them off in lengths similar to the appropriate elevation lines.

PATTERN FOR ROUND PIPE CUT OBLIQUELY

The following procedure should be used to develop the pattern on sheet metal. First draw the elevation 0-A-G-6, as at Fig. 2 (*a*), and on the base edge describe a semicircle to represent a half-section of the base. Divide the semicircle into six equal parts, letter the points B, C, D, E, and F, and draw vertical lines from them to meet the top edge line 0-6. These lines are shown dotted to the base line. Next draw a

horizontal line of indefinite length for the base or stretch-out
line of the pattern, mark a point A and measure off from it the

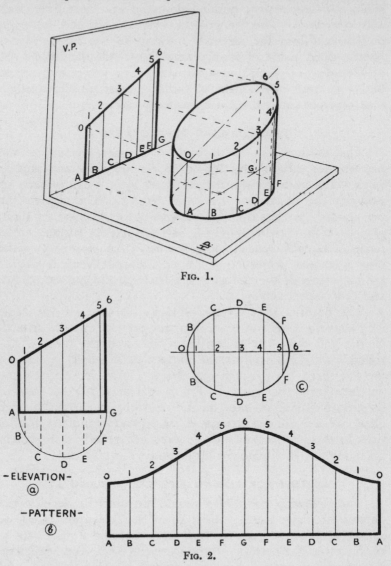

FIG. 1.

-ELEVATION-
ⓐ

-PATTERN-
ⓑ

FIG. 2.

circumference of the pipe.   This distance A-A is obtained by
multiplying the pipe diameter by 3⅐.   Divide A-A into twelve

equal parts and erect perpendiculars of indefinite length from
the points. Cut these lines off equal in length to the similar
full lines on the elevation surface, line 2-C, for example, in the
pattern, being the same length as line 2-C (the full line between
the top and base edges) in the elevation. Finally, draw a
smooth curve between the top points to complete the pattern.

The shape of the top section as at Fig. 2 (c) is found in
the following manner. Draw a horizontal line in a convenient
position, mark a point 0 and measure off distances 0-1, 1-2, 2-3,
3-4, 4-5, and 5-6 from the elevation top edge. Draw perpendiculars of indefinite length through the points, and cut
them off each side of 0-6 equal in length to the " ordinates,"
i.e., the dotted lines drawn from points B, C, D, E, and F to
the base line, in the elevation. Through all the points found
draw a smooth curve.

PATTERN FOR TEE-PIECE OF EQUAL DIAMETER

The example shown in Fig. 3 is of a cylindrical tee-piece
made of equal diameter pipes. To obtain the patterns for

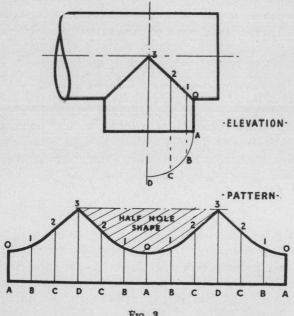

FIG. 3.

the job, first draw an elevation as shown and describe a quadrant on the base line. Divide it into three equal parts, A-B, B-C, and C-D, and draw perpendiculars to the base line from the points. These lines are shown dotted. Next produce the lines to meet the joint line between the pipes at points 1, 2, and 3. Draw a horizontal line of indefinite length in a suitable position and mark a point A. Calculate the circumference of the cylinder and measure it along the stretch-out line. Divide A-A into twelve equal parts and erect perpendiculars of indefinite length from the points. Letter these as shown. Cut the perpendiculars off equal in length to the similarly marked lines in the elevation and draw a smooth curve through the points found. The shaded portion of the pattern represents half of the shape of the hole in the main pipe.

RADIAL-LINE DEVELOPMENT

The radial-line method of pattern development is used for those objects, such as cones and pyramids, the sides of which converge to an apex. By far the most common object which requires marking out in the workshop is the right cone frustum. The method of drafting the pattern consists in using the slanting edge line of the full cone as a generator and describing a girth line which is made equal in length to the circumference of the cone base. Each of the free ends of the girth line are joined by straight lines to the cone apex, and from the same point the top edge curve of the frustum is drawn across the pattern. Many jobs, such as breeches pieces, can often be made up of portions of right or oblique cones, but unless the parts are of short taper the best plan is to obtain their patterns by triangulation.

CONE PATTERN

In Fig. 4 is shown the method of drafting the pattern for a right cone frustum. Draw an elevation of the frustum as shown and produce the edge lines to an apex, marked X. For the pattern mark a point X in a suitable position, and with radius A-X from the elevation describe an arc of indefinite length. Mark a point A and measure the circumference of the cone base round the arc, preferably with a flexible steel rule. Join the points to X. Next take radius 0-X from the elevation and transfer to the pattern as shown.

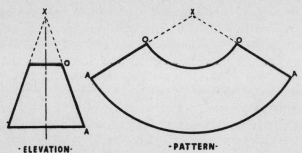

·ELEVATION·   ·PATTERN·

FIG. 4.

Fig. 5 illustrates another method generally used to obtain the length of the girth line. Describe a semicircle on the cone base line A-G and divide it into six equal parts, A-B, B-C, etc. Take one of these spacings and step it round the girth line twelve times. Join the end points to X and complete the pattern as for the previous example.

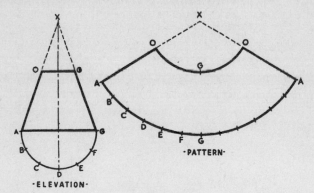

·ELEVATION·   ·PATTERN·

FIG. 5.

METHOD OF TRIANGULATION

The triangulation method is universally applied in the trade to solve a large number of development problems. In the main the standard system consists in drawing an elevation and plan of the object to be developed, dividing the surface of the plan into a suitable number of triangles, finding the true lengths of the triangle sides, and building up pattern triangles in the correct relation to the triangles already drawn in the plan. A plan of the object, that is, a view which shows the top or bottom edges of the material, is essential to this system.

Most working drawings give the front elevation, side elevation ("end view"), and the plan of the job. When triangulating the plan it is usually necessary to set-out this from the elevation, and in those cases where the top and base of the article to be developed do not lie between parallel planes, problems of projection are certain to arise. When an object, such as a ventilation transformer, has a top inclined at an angle to the base it becomes vital to draw a correctly projected plan from its elevation, which procedure calls for a fairly good knowledge of the principles of geometry.

Also, if the ends of any object are curved or undulating, as, for instance, the joint line formed by the intersection of two right cones, the projection of an accurate plan becomes a difficult proposition. Most workmen have experienced this trouble, particularly when jobs are of slight taper. The radial-line method of developing cannot then be used, as this system is only possible if an apex can be found. What usually happens in practice is that such patterns are made somewhere near to the desired shape by roughing-out methods and trimmed to size where necessary. This is a very unsatisfactory way of working and an alternative method is sometimes used, consisting of making models of the job and obtaining the dimensions of the pattern from them. Errors easily creep in by this method, and, as a rule, it takes too much time to be a practical proposition.

Because of this difficulty of developing satisfactory patterns the designing of much new work is restricted. In air-duct work, for example, there is great scope for the introduction of more efficient breeches pieces, transformers, etc., provided they can be made correctly in a reasonable time. To obviate most of these difficulties a new triangulation system, now to be described, was evolved. It is designed to meet the demands of the workman who requires an efficient method of pattern development which is simple in operation and easy to apply to any triangulation problem he is likely to encounter in everyday practice. In this system the projection of a plan from the elevation of the object is unnecessary, as all the true lengths are obtained directly from the elevation or end view, drawn in a right position.

ADVANTAGES OF LAY-OUT SYSTEM

Once the principle of the system is understood it is easily memorised ; in fact, if the few simple rules are strictly adhered to, a workman with the most elementary knowledge of geometry is able to mark-out patterns which he would have been unable to tackle by other methods. The developing of patterns for many jobs which involve interpenetration of surfaces are also brought within his capabilities, as such problems can often be more or less eliminated if desired. This is done by drawing an approximate joint line between the components of the job and the patterns triangulated, a correct fit between the respective parts when assembled being guaranteed. Other advantages of the system, apart from its simplicity of operation, are that a minimum of lines are used, and, as there are no plan lengths to cross over each other, the possibilities of errors are reduced and the laying-out of the pattern on the sheet metal is straightforward and logical.

It is advisable when drafting patterns to letter and number the points, as this avoids confusion when building up the triangles. Familiarity with triangulation methods, however, will eventually enable the sheet metal worker to dispense with this procedure, as experience in laying-out work cultivates a keen eye for the correct unfolding of a pattern.

The examples chosen for explanation purposes are typical of many that are to be found in the sheet metal industry. Most of them deal with ventilation ducts, which are made in an unending variety of forms that afford splendid opportunities for the pattern drafter to exercise his skill.

Chapter Two

DESCRIPTION OF LAY-OUT SYSTEM

THE following description of the system is intended to show the principle as clearly as possible, and at its conclusion a few simple rules will be given to enable the student to use the system in actual practice. To triangulate the pattern for an article, it is first of all necessary to draw its elevation in a right position. The surface of the elevation is then divided up into a convenient number of triangles by means of " false " length lines, and their true lengths

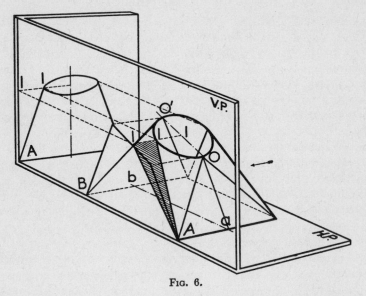

FIG. 6.

determined, to enable the pattern triangles to be built up. The method of obtaining these true lengths is illustrated in Fig. 6, which shows a pictorial view of a ventilation transformer, the round top of which is inclined to its square base. It is placed on a horizontal plane with one edge of its base touching the vertical plane, which must be assumed to be transparent to enable the elevation to be seen drawn upon it. The elevation view is projected from the outline of the transformer looking in the direction of the horizontal arrow, whilst the end view is projected as shown.

On the surface of the transformer is drawn the true length 1-A between the points 1 and A. This line is projected on to the elevation and end view, and as it is inclined to both planes it appears in the elevation and end view as a false length. Thus an elevation or end view of the transformer on a blue print would only show 1-A as a false length. If line 1-A be closely observed on the transformer it will be seen that it forms the hypotenuse of a right-angled triangle (shown shaded), the elevation or false length making the opposite side, the third side being the length of the projector 1-1, which connects the top edge of the transformer and the vertical plane. From this it should be obvious that, given the elevation length, the true length is determined by the length of the projector in any right-angled triangle. The projector line is, of course, always at right angles to the vertical plane. It can be seen from the end view that the length of the projector is the difference in length between the centre line of the transformer top and the vertical plane—which is actually the half-width of the base—and the half-width of the top.

The above principle, which the pictorial view should have now made clear, is applied in practice in the following manner. A drawing of the job is shown in Fig. 7, the false length line 1-A being drawn in the elevation. The position of 1-A is determined by describing a semicircle, representing a half-view of the top circular face, on the top edge line, and drawing an ordinate from the semicircle to the edge line, marking it 1-1. The end view has a line A-A drawn parallel to its centre line ; this line represents the edge view of the vertical plane and the line between it and the centre of the transformer top is clearly shown. A triangle 1-A-1 is thus formed between the vertical plane and the transformer and is the end view of the shaded triangle in the pictorial view. To make this triangle its true length, point 1 on the top edge is projected down to the base line marking 1 and the elevation false length 1-A extended along A-A from the corner of the base, also marking 1. This point is joined to 1 on the base line,

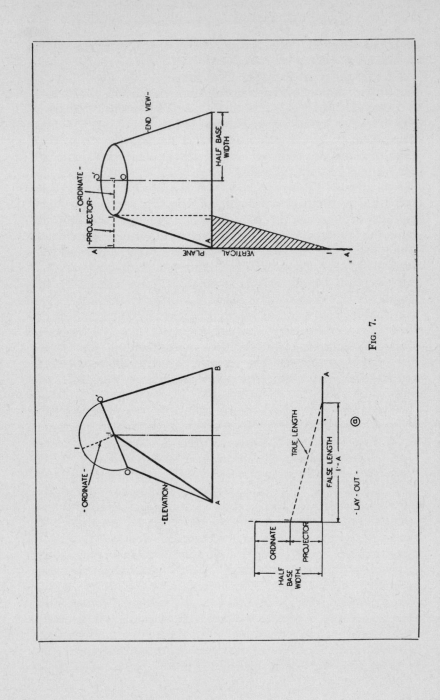

Fig. 7.

thus obtaining the true length of 1-A and completing the triangle.

Using this principle, the true length of any pattern line is deduced in a lay-out from its apparent length in the elevation. To construct the lay-out, a vertical line of indefinite length is drawn in any convenient position, Fig. 7 (*a*), and a point marked on it. Number this point 1, to denote the centre of the top of the job. The half-width of the base is next measured off from 1 and an indefinite horizontal line drawn from the point found. From 1 is also marked off the length of the ordinate 1-1 from the elevation. The distance between this point and the corner of the right angle formed is the projector length. The false length 1-A from the elevation is marked off along the horizontal line from the corner of the angle, and the point marked is joined to the nearest point 1 on the " perpendicular," thus obtaining the true length required.

It is not necessary to repeat all this construction of the lay-out for each false length line from the elevation, as it is quite correct to use the same right angle for any number of lines. To do this, adopt the following procedure for the lay-out when triangulating the patterns for any articles of which the bases are square or rectangular :—

1. Draw a vertical line and mark off on it a point to denote a top centre line of the object.

2. From this point measure down the top ordinates or half-widths and number the points.

3. From the same point measure down the half-width of the base and draw a line of indefinite length from the point found, at right angles to the vertical line.

4. Step off, from the corner of the angle formed, the elevation false length lines along the horizontal line.

5. Join the points marked off on the horizontal line to the numbered points on the perpendicular in the correct order, so obtaining the true lengths of the distances measured on the horizontal line.

The pattern triangles are then built up with the true lengths in a manner to be explained later.

PATTERN FOR TRANSFORMER

From this explanation of the system there should be little difficulty in applying it to the development of the transformer pattern, using the elevation only. It is only necessary to develop a half-pattern for the transformer, as the drawing shows that each half is symmetrical about the end view centre line. The full pattern can be drawn as desired.

Draw the elevation as in Fig. 8, lettering the base corners A*a* and B*b*. On the edge line 0-6 describe a semicircle, divide it up into six equal parts, and draw perpendiculars to 0-6. Number each ordinate 1, 2, 3, 4, and 5 respectively. Connect the points on 0-6 to the base corners A*a* and B*b* with full lines, 1, 2, and 3 to A*a* and 3, 4, and 5 to B*b*. The elevation surface is thus divided up into triangles, each connecting line between the top and base edges being a " false " length. Each of the full lines is designated by its appropriate ordinate number, line 2-A, for instance, in Fig. 8 being the distance between the base point of ordinate 2 on the top edge 0-6 and the corner point A*a*. The edge lines of the elevation 0-A*a* and 6-B*b* are both true and false lengths. This will be made clear by reference to the pictorial view in Fig. 6. Consider the line 0-*a* on the transformer. This line is parallel to the vertical plane, therefore it appears as a true edge line in the elevation. At the same time the edge line is also the projected or false length of 0-A on the transformer. This is the reason the edge lines in the elevation are lettered A*a* and B*b* at the base corners (Fig. 8).

The lay-out is next drawn in a convenient position. In accordance with the rules previously set out, a vertical line is drawn, point 0 and 6 marked on it, and the ordinate lengths measured off from 0 and 6 and numbered as shown. Each point, it will be noted, has two numbers, because the lengths of 1 and 2 ordinates are the same as 5 and 4 respectively. The half-width of the square base is next measured off from point 0 and 6 and a line drawn at right angles to the perpendicular and lettered A and B for each corner.

To start the half-pattern, draw in a suitable position the edge line of the elevation and mark it 0-*a*. Next mark off the same line along the horizontal line of the lay-out and

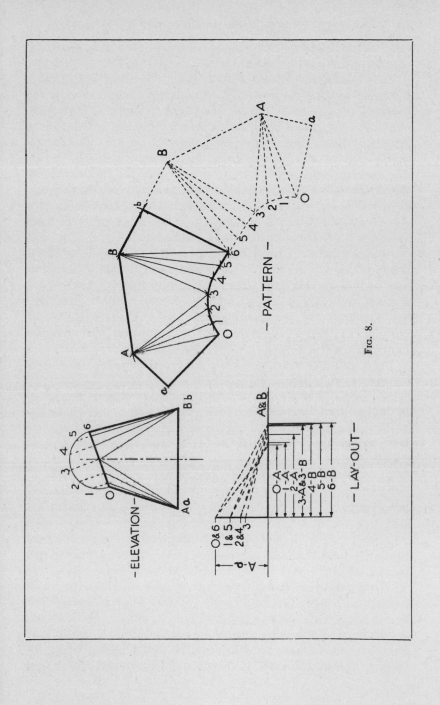

– PATTERN –

– ELEVATION –

– LAY-OUT –

O-A
1-A
2-A
3-A & 3-B
4-B
5-B
6-B

A & B

O & 6
1 & 5
2 & 4
3

d
A

Fig. 8.

join the point marked to 0 on the perpendicular. With this true length describe an arc from 0 on the pattern lay-out to the right of *a*. Take the half-base width A-*a* and, using *a* as centre, cut the arc previously drawn, at A. Join *a* to A and 0 to A to complete the first pattern triangle. It is very important to note that the number of the false length line determines to which point it is connected on the lay-out perpendicular, 0-A joining to 0, 1-A joining to 1, and so on. The best method is to pick up a line, 0-A for instance, with the dividers, mark it off on the horizontal of the lay-out, keep one point of the dividers on the point marked and open the dividers to place the other point on the correct number on the perpendicular, which is 0.

To continue the development, take the false length 1-A from the elevation—the full line—measure it along the base line of the lay-out, and from the point marked extend the dividers to point 1 on the perpendicular. With this true length describe an arc to the right of 0 in the pattern lay-out, using A as centre. Cut this arc with spacing 0-1 from the semicircle in the elevation, using 0 as centre.

Next take distance 2-A from the elevation, measure it off on the lay-out base line and extend to point 2 on the perpendicular. Describe an arc with the true length from A in the half-pattern and cut this in 2 with spacing 1-2 from centre 1. Take false length 3-A, obtain its true length in the lay-out, and cut an arc from centre A in the half-pattern in 3, with spacing 2-3.

The next line, 3-B, is the same length as 3-A, so strike an arc from 3 with this distance and cut it in B with the true base length A*a*-B*b* from the elevation. Join A-B. Take false length 4-B next and measure it along the lay-out base line. Triangulate to 4 on the perpendicular and strike an arc from B in the half pattern. Cut this arc in 4 with spacing 3-4, using 3 as centre.

Take false length 5-B, measure off and triangulate in the lay-out, and with the true length describe an arc from B and cut it in 5 with spacing 4-5. Take the edge line 6-B*b*, triangulate in the lay-out, and swing an arc from B in the half-pattern with the true length. Cut this in 6 with spacing 5-6 from centre 5. Finally, take the edge line as a true length from the elevation and with it describe an arc from 6, then cut this arc in b with B-b, which being the half-width of the base is equal to A-a. Join B to b, and b to 6.

Draw a smooth curve through the top points on the half-pattern. To complete the full pattern it is necessary to reproduce the lines already drawn. This is done by first picking up line 6-B in the half-pattern, swinging an arc from 6 to the right of the edge line 6-b, and cutting it in B with the half-base width B-b from centre b. From B an arc is struck to the right of 6-b with distance 5-B, and it is cut in 5 with spacing 5-6. Continue the development in this manner; taking each line in turn from the half-pattern will enable the whole job to be set-out in a very short space of time.

The student would be well advised to go over the foregoing explanation very carefully with a pair of dividers in hand and check up the description of the development with the lines shown in the diagram. In this way he will become familiar with the principle of the system, and he can then proceed to draft the pattern on paper. It is a good plan to cut out the finished development after allowing a small overlap on one of the edge lines to enable the joint to be gummed together when the model is shaped up. With practice the student should soon be able to use the system in the workshop, marking out the job on sheet metal, and in this way gaining sufficient confidence to tackle any of the jobs to be described later.

For extra practice, patterns for similar types of transformers can now be attempted. The pattern for a transformer with a rectangular base could be drafted, the only difference from the one just developed being that the base distance Aa-Bb should be drawn longer. Also, the development by similar means of a square to round transformer between parallel planes may be drafted, a quarter-pattern only being necessary if the top and base are drawn on common centre lines. Other types of transformers will be described in later chapters.

No allowances for grooved, riveted, or soldered joints are made on any of the patterns in this work. Such allowances can readily be added where necessary, although as many jobs are butt welded nowadays, the net pattern is usually all that is required.

Chapter Three

FURTHER DESCRIPTION OF LAY-OUT SYSTEM

THE new triangulation system is easily applied to the development of patterns for articles which have square or rectangular bases, but, of course, there are innumerable jobs which differ from these conditions. It must be clearly understood that the shape and contour of the ends, faces or sections of any article to be triangulated, or the angle at which they lie to each other, do not in the least affect the use of the lay-out system. Once the elevation or end view of an object is drawn and the correct shape of its top and bottom faces determined, it becomes merely a routine matter of triangulating the surface of the elevation and deriving the true lengths on the lay-out. In this way all the difficulties of triangulation disappear, because there are no plan lengths with which to contend.

Practically all jobs can be triangulated from the elevation as drawn on a working drawing or blue print, as it is unnecessary to draw other elevation or plan views to enable true sections to be obtained in their correct relation to each other. If the projected plan of an object can be simply drawn as, for instance, two concentric circles representing the frustum of a right cone, its pattern can be obtained by the usual methods of triangulation if desired ; but if a plan is required that calls for much projection work from the elevation, the lay-out system should be used.

To triangulate the pattern for a ventilating connecting piece, as shown in the three views of it in Fig. 9, it is necessary to understand certain modifications to the lay-out to suit the altered conditions from those previously described. It will be seen that the job has a circular top inclined to a circular base of larger area, both being symmetrical on a common centre line. In the elevation, plan, and end view, is drawn a false length line 1-B, the position of which is determined by the ordinates 1 and B on the top and base of the elevation.

The principle of obtaining its true length from the elevation is as follows :—

An edge view of a vertical plane D-D is shown in the end view. The false length 1-B is projected to the vertical

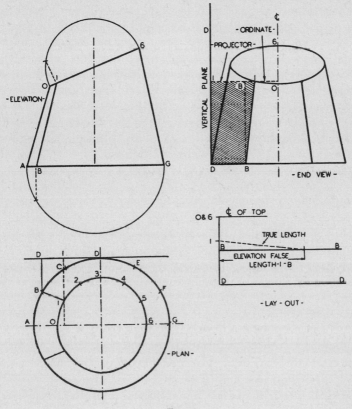

FIG. 9.

plane from point 1 on the top edge of the job and from B to D on the base line. This means that there are two pro-jector lines, 1-1 and B-D. Instead of the triangle as formed in the end view in Fig. 7, there is now a trapezoid 1-1-B-D. If, however, point B is projected up to 1-1, marking B, a triangle B-B-1 is formed. The distance from B to 1 on the top edge is thus the difference in the lengths of the two projectors 1-1 and B-D.

This procedure gives a method of drawing a lay-out to obtain the true length of 1-B. Draw a vertical line as shown and mark a point 0 and 6 on it to represent the centre line of the top. Measure down from this point the half-diameter of the base and mark the point D. Also from 0 and 6 measure the same distance as between B and the centre line of the end view, which is actually the base ordinate B in the elevation. Mark this point on the perpendicular B. Draw lines of indefinite length from 0 and 6, B and D at right angles to the perpendicular. Next measure on the perpendicular from 0 and 6 the top ordinate 1, and mark the point accordingly. Take the elevation false length 1-B and mark it off from B on the lay-out along the horizontal line, and from the point marked join a line to 1 on the perpendicular to obtain the true length. In this manner there is reproduced the triangle B-B-1 in the end view elongated to the length of 1-B in the elevation. It should now be obvious that the deciding factor in determining a true length is *the difference in the lengths of the projectors* between points on the edges of the object and a vertical plane contiguous to it. This is the principle of the lay-out, in which the projector lengths are fixed between a location point derived from the top of the object and a base line representing the position of the vertical plane.

In actual practice the lay-out is drawn as follows :—

1. Draw a vertical line and mark off on it a point to denote a top centre line of the object.

2. From this point measure down the top ordinates or half-widths and *number* the points.

3. From the same point measure down the half-base widths or ordinates and from each point draw a line of indefinite length at right angles to the vertical line, marking its appropriate *letter* on each line.

4. Step off each false length on the horizontal line marked with the same letter, measuring from the corner of the angle formed by the vertical and horizontal lines.

5. Join the points marked off to the appropriate *numbered* points on the perpendicular, so obtaining the true lengths required.

SETTING-OUT FOR CONNECTING PIECE

By following the above rules the student should now be able to develop the pattern for the connecting piece as shown

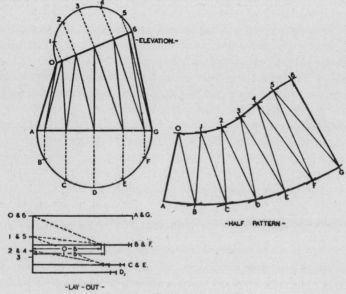

FIG. 10.

in Figs. 9 and 10. Draw the elevation 0-6-G-A, and describe a semicircle, representing the half-surface of the top, on 0-6, divide into six equal parts and number 1, 2, 3, 4, and 5, as shown. Next draw perpendiculars to the top edge, so obtaining the ordinates for the lay-out. On A-G describe a semicircle which is half the outline of the base and divide it into six equal parts, marking the points B, C, D, E, and F. Draw in the ordinates to the base edge A-G and join up all the points as shown, thus dividing the surface of the elevation into triangles.

The lay-out is next drawn in a suitable position, the top ordinate points being measured off from 0 and 6 and numbered 1 and 5, 2 and 4 and 3. From the same point is marked off the base ordinates and parallel lines drawn from the points, marked with the appropriate letters from the base semicircle.

It is only necessary to draft a half-pattern, because both halves of the job are the same each side of the horizontal centre line in plan. Start by drawing in a suitable position the edge line of the elevation 0-A. Next take the false length 0-B, mark it off on B line on the lay-out and join the point marked to 0 on the perpendicular. With this true length describe an arc from 0 on the line 0-A to the right of A. Take the distance A-B from the base semicircle, and, with A as centre, cut the arc previously drawn in B. Next pick up the false length 1-B, mark off on B line on the lay-out, join to 1, and with this true length describe an arc from B in the pattern lay-out. Cut this arc from 0 with distance 0-1 from the semicircle on the top edge, so obtaining point 1.

The false length 1-C is next placed on C line on the lay-out, and the point marked joined to 1. This length is then added in the same manner as the others to the half-pattern. The remainder of the triangles are dealt with in similar fashion until the half-pattern is complete. Join all the points on the top and base of the pattern with a smooth curve after reproducing the triangles for the full pattern.

TAPERED LOBSTER-BACK BEND

A problem which can be very easily solved by the lay-out system is that of the tapered lobster-back bend. In this job each of the segments are circular ended, the segments tapering gradually from a large to a smaller diameter pipe. As a rule the joints between the segments are butted together and welded, so there is no necessity to make metal thickness allowances for slip-in joints, the circumference of the small end of the base segment being exactly the same as the large end of the next segment, and so on. An elevation of the lobster-back is shown in Fig. 11, the base-edge line A-G being the diameter of the large pipe and Y-Z the diameter of the reduced end. The elevation is divided up into the number of segments required, and the patterns for each one have to be got out separately, the method being, however, the same in each case.

To develop the half-pattern for the base segment A-0-6-G,

it is essential to divide it up into a convenient number of triangles and find their true lengths on the lay-out. Draw a semicircle on A-G, as shown, to represent half the circumference of the large end of the segment. Divide the semicircle up into six equal parts, number B, C, D, E, and F, and draw perpendiculars (shown dotted) to the base line A-G. Next draw a semicircle on 0-6, divide this also into six equal parts, and draw perpendiculars to the top-edge line. Join all the points with false length lines to obtain the elevation triangles.

The lay-out is next drawn according to the rules previously set out. On the perpendicular a point is marked 0 and 6, and the top ordinates marked from it and numbered. The bottom ordinates are measured off from the same point, horizontal lines drawn and lettered the same as those on the base semicircle. The true lengths of the elevation triangles are found on the lay-out similar to the previous example, and the half-pattern developed in the same way.

It will readily be seen that the middle segment has for its *base* line the distance 0-6, and the surface of it can be divided up into triangles by drawing a semicircle on its top edge, placing in the ordinates and joining the points to the points on 0-6. To save confusion with the lay-out for this segment delete the numbers on the ordinates on 0-6 and mark them *b, c, d, e,* and *f,* as shown. Construct the lay-out in the usual manner and develop the pattern triangles from it. The remaining segment is dealt with similarly. By these means the patterns for the whole job can be quickly and accurately drawn, as the bugbear of projecting the plan of each segment with the consequent confusion of the plan length lines, particularly in the segment throats, is entirely eliminated. In addition, all the spacings for the circumferential joints are obviously taken from the elevation, and therefore the possibility of errors, by using their elliptical plan outlines, cannot arise.

CONE FRUSTUM PATTERN

One of the most common objects which requires to be set-out is the pattern for a frustum of a right cone. In Fig. 12 is shown an elevation of such a job, the pattern for which is drafted as follows. First set-out the elevation and describe a semicircle on the top-edge line. Divide this up into equal

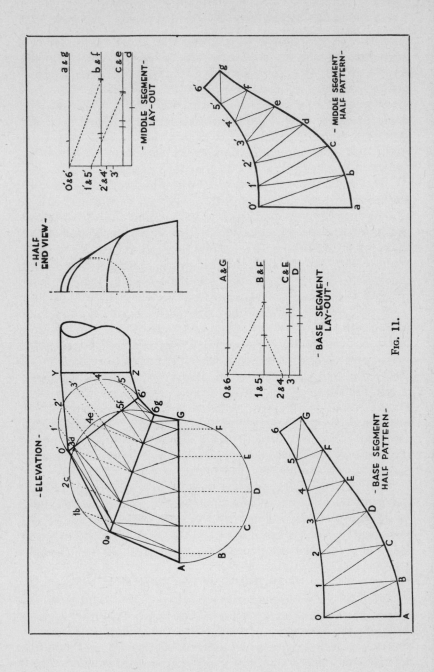

- HALF END VIEW -

- ELEVATION -

- MIDDLE SEGMENT LAY-OUT -

c - MIDDLE SEGMENT HALF PATTERN -

- BASE SEGMENT LAY-OUT -

- BASE SEGMENT HALF PATTERN -

Fig. 11.

parts, 0-1, 1-2, and 2-3. Next describe a semicircle on the
base edge, divide up as shown, and draw an ordinate from
point B to the edge line. Join the point on the base edge to 0
with a false length line.

Draw two lines of indefinite length mutually perpendicular
to each other for a lay-out. Cut the vertical line 0-X off
equal in length to ordinate B in the elevation. Take the false
length 0-B, measure it off on the lay-out base line from the
corner of the angle. From the point found join a line to 0
on the perpendicular to obtain the true length line for the
pattern development.

Start the pattern by drawing in a suitable position a line
equal in length to the edge line 0-A in the elevation. This is

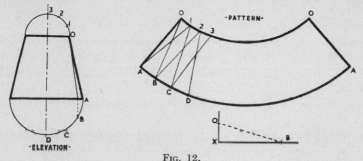

Fig. 12.

the seam line for the pattern. Next pick up with the dividers
the true length 0-B from the lay-out. Describe an arc from
0 on the seam with this true length and cut it in B with spacing
A-B from the semicircle, using A as centre. From A in the
pattern lay-out describe an arc with the same true length
line and cut this in 1 with spacing 0-1 from the elevation
top edge. From point 1 describe an arc to the right of B
with the true length line. Cut this arc in C with the spacing
B-C. Next, from point B, swing an arc with the true length
and cut it in 2 with spacing 1-2.

Continue the development on the same lines throughout
until the full pattern is complete. One method by means of
which the pattern can be quickly and neatly drafted is to use
three pairs of dividers. Set one pair to the length of the true
line and the other pairs to the length of a top edge and bottom
edge spacing respectively, and pick them up in turn as required.

Chapter Four

JUNCTION PIECES

PERHAPS the most interesting of the many problems met with in air-duct and dust-extraction work are those which are concerned with the development of patterns for breeches and multiple-way junction pieces. These jobs are made in a number of different ways, depending on the particular design of the plant, the most common type having two or three circular-ended branches converging into a round main.

Breeches pieces are designed to distribute or converge the air flow with as little friction as possible. It is often advocated that these jobs should be made up of portions of oblique cones to facilitate the drafting of their patterns by the radial-line method. This method has certain disadvantages, particularly for large work, as it is difficult to produce the edge lines of the branches to a radius point within a reasonable distance. Poor design is also sometimes responsible for having the inlets too far apart, thus reducing the branch cross-section area. In most cases the joint face is best made semi-elliptical, to keep the shape of the branches the correct area and to give a pleasing appearance. This design necessitates the use of triangulation to obtain the patterns, and the lay-out system is particularly suitable for setting-out this class of work because it is unnecessary to determine the geometrical solid of which the branches are made up, as their surfaces are triangulated between suitably shaped end and joint faces.

An end view or elevation of the job used in conjunction with a lay-out are all that need to be drawn to obtain the true length lines for the pattern triangles. True length lines are obtained without the use of a projected plan in every case. It is not necessary to find the exact position of a joint line between the branches of a junction piece by determining points on their surface interpenetrations, as a suitable joint line can be drawn, within certain limits, to a desired contour. Each of the branches is then triangulated to the shape drawn, so ensuring the same length of joint line for each part. This procedure eliminates a good deal of the geometrical construction that is required when developing multiple-way piece patterns, particularly when the branches vary in size and shape.

The simplest type of breeches piece has branches similar in shape, as they transform from circular inlets to a circle, the area of which is equal to the combined areas of the inlets.

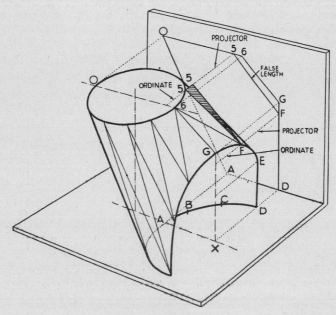

FIG. 13.

Fig. 13 shows a perspective view of one branch of this type of breeches, placed on a horizontal plane. On the vertical plane is depicted an elevation as projected from the outline of the object. The face of the joint cut between the two branches is clearly shown, it being semi-elliptical in shape, the diameter of the base forming the minor axis, whilst half the major axis is the distance between the base at X and G.

Points G, F, and D are shown projected from the joint line to the elevation. The method of obtaining the true lengths of lines on the surface of the object, drawn between points on the top edge and points on the joint line, is the same as for

previous problems. True length line 5-F on the branch,
for instance, is projected from points 5 and F on to the elevation,
in which it appears as a false length. The difference in the
length of the projector from 5 on the top edge to 5 on the
elevation and the projector from F to F is made apparent by
the line 5-F drawn from F, parallel to the vertical plane.
This line forms the base of the shaded triangle 5-F-5 and its
position in the lay-out is determined by measuring the length
of the ordinate F from a point on the lay-out perpendicular
denoting the centre of the top of the object.

PATTERN FOR BREECHES PIECE

The method of developing the pattern for the branch is
shown in Fig. 14. Draw an elevation A-0-6-G-X and pro-

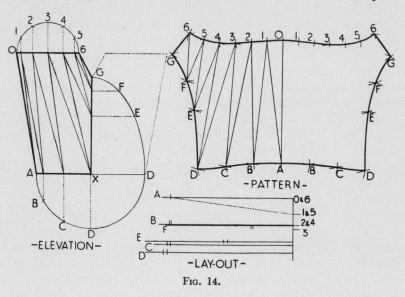

Fig. 14.

duce the base line to D. Next describe a semicircle on the top
edge and divide it up into six equal parts. Number the points
0, 1, 2, 3, etc., as shown. Drop perpendiculars to the top
edge, so obtaining the ordinates for the lay-out. Describe
a semicircle on the base edge, divide one-half into three equal
parts and mark the points A, B, C, and D. Draw in the
ordinates to the base line.

The shape of half the joint cut is next drawn on X-G. A quarter-ellipse is drawn between points D and G and points E and F, marked in at suitable positions on the curve. Place in the ordinates to the joint line X-G, as shown. Divide up the elevation surface into triangles by connecting the points on the edges by means of false length lines. The true lengths of the elevation triangles are obtained on a lay-out, which is constructed as follows. Erect a vertical line of indefinite length in a convenient position and mark off a point, 0, on it. Measure down from 0 the lengths of the ordinates from the elevation top edge, numbering them 1 and 5, 2 and 4, and 3. Also from 0 mark off the lengths of the ordinates from the base and joint face. Draw lines of indefinite length from the points found and mark them A, B, C, D, E, and F respectively. Note that line A is drawn from 0, because the breeches is symmetrical about its plan horizontal centre line, so points 0 and A are on the same plane. The base line, D, of the lay-out actually represents the position of the vertical plane in its correct relation to the horizontal centre line.

To start the pattern, draw a line 0-A in a suitable position. This is the edge line 0-A in the elevation, and it is made the middle line of the pattern to ensure that the seam will be at the smallest part of the branch. Now pick up with the dividers the false length 1-A from the elevation and mark it along line A in the lay-out, measuring from 0. From the point marked draw a line (shown dotted) to 1 on the perpendicular, to obtain the true length of 1-A. Add this line to 0-A in the pattern lay-out by describing with it arcs each side of 0, using A as centre. Cut these arcs from 0 with the spacing 0-1 from the top edge semicircle. Next take the false length 1-B, mark it off along line B in the lay-out, from the corner of the angle, and join to 1 on the perpendicular. It is not really necessary to draw a line between the points, as the true length distance is picked up with the dividers. With the true length 1-B describe arcs each side of A in the pattern lay-out, using 1 as centre. Cut these arcs in B with the spacing A-B from the base semicircle. Pick up the false length 2-B from the elevation, step it along B in the lay-out and triangulate to 2 on the perpendicular. Strike arcs with the true length from points B in the pattern and cut them in point 2 with distance 1-2.

Each half of the pattern is built up on the same principle, care being taken to step off each false length in turn on its appropriate lay-out line and triangulating to the correct number on the perpendicular. To add line 2-C to the pattern, pick up the elevation false length, step it off on lay-out line C, and extend to 2 on the perpendicular. Describe arcs from point 2 in the pattern and cut them in C with spacing B-C. Obtain lines 3-C, 3-D, and 4-D in similar fashion. For pattern line 4-E take the false length from the elevation, mark it off on lay-out line E, and triangulate to 4 on the perpendicular.

From centre 4 in the pattern describe an arc with the true length and cut it in E with spacing D-E from the quarter-ellipse in the elevation. Take false length E-5, obtain its true length in lay-out, swing an arc from E in the pattern and cut it in 5 with spacing 4-5. Continue the remainder of the development similarly, using spacing E-F and F-G from the elevation, as shown. The edge or seam line G-6 is a true length, and as such is picked up direct from the elevation. Finally, draw an even curve through all the base and top edge points.

MULTIPLE JUNCTION PIECES

Fig. 15 shows one branch of a three-way piece, placed on a horizontal plane in such a position that its centre line 0-6 is parallel to the vertical plane. The round inlet is inclined to the base, which is divided up into three equal sectors. It can be seen that the face of the joint cut is quarter-elliptical in shape, and the projected view of this in the elevation is shown by the line drawn from C to G. Two more similar branches can be fitted together to form the three-way, but it must be clearly understood that any other shape of branch can be inserted into the sectors if desired, providing, of course, that one adheres to the same shape of base and joint contour from C to G. In this manner one of the branches could have its inlet parallel to the base, whilst another branch could have a square inlet, and so on.

The elevation of each branch is drawn separately with the correctly projected shape of the joint line placed in. Junction pieces with four or more branches are made in practically the same manner, the only difference being that the base is divided up into the required number of sectors, with a consequent difference in the contour and position of the elevation joint curve. Branches made on this principle are easily and rapidly put together. It is a good plan when shaping-up to use a template cut to the contour of the joint curve, each branch

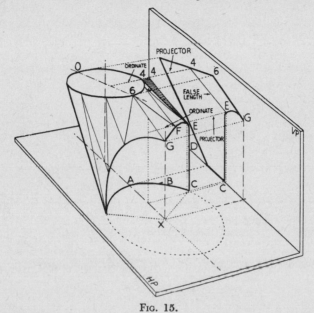

FIG. 15.

being then shaped to it. This process enables the parts to be assembled with a minimum of trouble. There is no need in every case to make the joint curve semi-elliptical, as deviations from this shape can be used. A good appearance, for example, can often be given to the branches by making the curve at F (Fig. 15) much more full and higher than point G. The principle of obtaining true length lines is the same as for previous problems, as a study of the perspective view will make clear.

PATTERN FOR THREE-WAY JUNCTION PIECE

To obtain the pattern, as shown in Fig. 16, draw an elevation, A-0-6-G-X, of one branch. On the base line describe a semi-circle representing half of the base outline. As the base of the junction piece is divided up into three parts it is necessary to draw in the sector A-X-C, as depicted in Fig. 15. To do this,

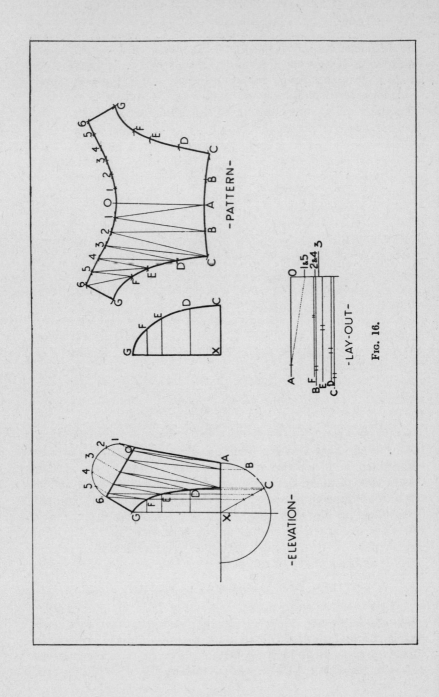

-PATTERN-

-LAY-OUT-

-ELEVATION-

Fig. 16.

divide a quarter of the base outline into three equal parts, mark B and C and join C to X. Now draw in ordinates from B and C to the base line. Adjacent to the elevation next construct the true shape of the joint face by erecting a line parallel to X-G, cutting it off the same length, and drawing X-C at right angles to it. Join C to G by an elliptical curve and divide it up into suitably spaced parts C-D, D-E, E-F, and F-G. Next draw ordinates to X-G. From X in the elevation measure off these ordinates along X-C from X, and from the points found erect lines of indefinite length parallel to X-G. The lines are shown dotted up to the base line, and these distances are used for the lay-out construction. Draw lines from D, E, and F on the joint face parallel to the base, to the elevation line X-G. Where these lines intersect those drawn through the elevation from X-C gives points D, E, and F. Draw a smooth curve through the points. Next describe a semicircle on the elevation top edge, divide it up into six equal parts, number the points 0, 1, 2, etc., and place in the ordinates. Join the points on the edges and joint line of the elevation by means of false length lines, so obtaining the surface triangles. Construct the lay-out by drawing a vertical line, marking point 0, measuring down the top edge ordinates and numbering them as shown. Take the lengths of the dotted lines between X-C and X-A and mark them off from 0 on the lay-out perpendicular. Mark these lines D, E, and F, also place in the lay-out lines A, B, and C. It is very important to note that the widths between the line X-C and the base line be used for the lay-out and not the ordinates from the quarter-ellipse.

The pattern is developed in the same manner as for the breeches piece, each false length from the elevation being triangulated on the lay-out and the true lengths used to build up the pattern triangles. Take the spacings 0-1, 1-2, etc., from the top edge semicircle for the top line of the pattern and the four spacings A-B and B-C from the base semicircle. Also take the distances C-D, D-E, E-F, and F-G in the pattern from the true joint curve, as they must not be confused with the spaces on the elevation joint line. After all the triangles have been constructed, join up all the points with a smooth curve to complete the pattern.

Chapter Five

SPECIAL TYPES OF JUNCTION PIECES

THERE are occasions when it is necessary to make an air-duct breeches piece with a flush side to enable the job to lie flat against a wall or bulkhead. This means that the inlets are off-centre from the base of the breeches, and consequently this difference in design from previous examples has to be taken into consideration when constructing the layout used for developing the patterns. In Fig. 17 is shown the elevation and end view of a flush-sided breeches, the circular inlets of which lie at an angle to its round base. As both branches are identical in shape and size the triangulation of the elevation surface of one branch only is required. It is not essential to draw a full end view for developing purposes, as all that is required is the shape of the joint curve, drawn adjacent to the elevation.

FLUSH-SIDED BREECHES PIECE

The pattern is obtained by the following means. First draw the elevation of one branch, as depicted in Fig. 17, and describe a semicircle on the top edge. Divide it up into six equal parts, 0-1, 1-2, etc., and draw in ordinates from the points to the edge line. Through the base line also describe a semicircle, divide into six equal parts, mark the corner A and the points below the base line, B, C, and D. The points above the base line are next marked *b*, *c*, and *d*. Join B to *b* and C to *c*, thus obtaining points on the base edge for the elevation triangles. Mark off ordinate 3 from the top edge, from D on D-*d*, and draw a line 0-6 through 3, parallel to the base edge. This line represents the position of 0-6 in its relation to the base, and it is used as a datum line to obtain points for the lay-out. On an extension of the elevation base line construct the shape of the joint face between the branches by marking off D-*d* from the elevation. Next measure D-3 from the base semicircle from D, and erect a perpendicular from 3. Mark off the height of the joint line at G and draw

40

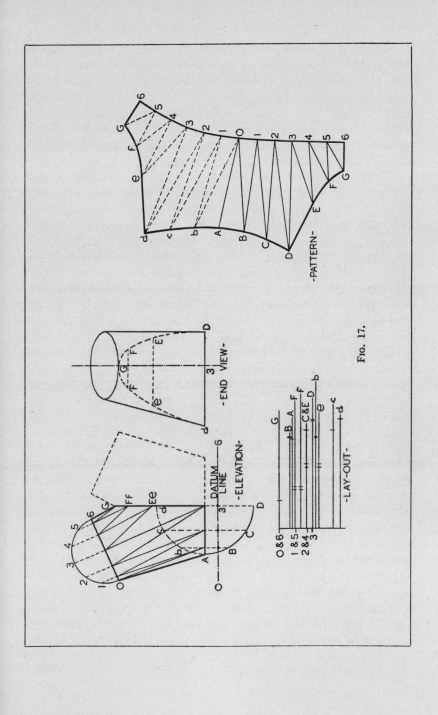

-PATTERN-

-END VIEW-

-ELEVATION-

DATUM
LINE

-LAY-OUT-

O & 6
1 & 5
2 & 4
3

Fig. 17.

in a suitable joint curve through G from points D and *d*. Divide the curve into a convenient number of spaces D-E, E-F, etc., and join by parallel lines points F and *f* and E and *e*. Project points *e* and *f* across to the elevation joint edge and mark as shown. Join all the edge points by lines to form the elevation triangles. Note that the base edge point determined by B-*b* is joined to 0 on the top edge, as this is essential to the correct development of the pattern.

To construct the lay-out for obtaining the true length pattern lines, draw a vertical line of indefinite length and mark on it a point 0 and 6. Step off from 0 and 6 the top ordinates, marking them 1 and 5, 2 and 4, and 3. From 3 draw a horizontal line of indefinite length and mark it D. Measure on the perpendicular from 0 and 6 the widths between B and C on the base semicircle and the datum line. Draw lines from the points, parallel to D, marking them B and C respectively. Next measure the width between A and the datum line and place line A in the lay-out. Repeat this procedure for lines *b*, *c*, and *d*, measuring them off from 0 and 6. These distances are those which lie between *b*, *c*, and *d* on the base semicircle and the datum line. Now measure the distances between line G-3 on the end view of the joint face, and E, F, *f*, and *e* respectively, and place these lines in their correct positions in the lay-out, as shown. Finally, draw G from 0 and 6. To be theoretically correct the pattern development calls for two lay-outs, as the positions of the projectors are being determined to contiguous vertical planes each side of the top centre line, but in practice the base and joint face widths are all marked down the perpendicular from 0 and 6. This simplifies the construction of the lay-out without any sacrifice of accuracy.

To start the pattern, pick up with the dividers line 0-A from the elevation. Mark this distance off along line A in the lay-out, measuring from the corner. From the point found extend the dividers to 0 on the perpendicular. Draw this true length 0-A in a suitable position, using it as the middle line of the pattern. Next take line 0-B from the elevation, measure it along B in the lay-out, and join to 0. With this length describe an arc from 0 in the pattern lay-out, and cut this arc in B with the spacing A-B from the elevation base.

Take elevation length 1-B, measure it along lay-out line B, and triangulate to 1 on the perpendicular. To add this true length to the first pattern triangle, describe an arc from B and cut it with the spacing 0-1 from the top-edge semicircle. Take line 1-C from the elevation, mark off along C in the lay-out, join to 1, and describe an arc from 1 in the pattern ; cut it in C with spacing B-C.

Repeating this procedure for the remainder of the elevation false lengths completes half of the pattern, using the distances D-E, E-F, and F-G from the joint curve in their correct positions in the pattern. Take the end line 6-G directly from the elevation, as it is a true length. Triangulate the remaining half of the pattern by picking up with the dividers the same elevation lines, but marking them off on b, c, d, e, and f lines respectively in the lay-out. For instance, take elevation length 0-b, measure it along lay-out line b, and extend to 0 on the perpendicular. Describe with this true length an arc from 0 in the pattern, cutting it in b with the arc A-b. Construct the remaining triangles similarly, taking care to place the elevation lengths on their appropriate lettered lines in the lay-out, and triangulating to the correct numbers on the perpendicular. The correct spacings from the joint curve must also be used, but careful consideration of the markings will prevent the possibility of errors. Complete the pattern by drawing smooth curves through all the points.

DESCRIPTION OF ALTERNATIVE LAY-OUT

The objects so far described have had their patterns developed from lines drawn on their elevation surfaces and made into true lengths in a lay-out. Certain types of jobs, however, lend themselves to easy development from their end views. The drawing of the flush-sided breeches, as depicted in Fig. 18, has two circular inlets parallel to its round base, and is a good example to which to apply this alternative method. It necessitates a slightly different type of lay-out, but it must be emphasised that the principle of obtaining true length lines remains the same. As previously explained, the lay-out is so constructed that the differences in the lengths of the projectors from an object to a contiguous vertical plane are fixed in it.

In Fig. 18 is shown the lay-out construction for use with

the end view. A vertical plane is assumed to divide the two branches of the breeches piece, its edge view being shown in the elevation. The top edge points on the elevation are marked 3 and 3' as shown, the centre being marked 0. D is marked on the base edge corner, whilst the top point of the joint line is lettered J. This line is extended below the elevation and also marked J. A line of indefinite length is drawn parallel to the base line of the elevation, at a convenient distance below it to J.

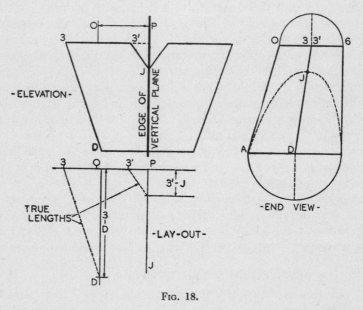

Fig. 18.

On this line is measured off the length of the projector from 3' to the vertical plane, marking point 3'. The points 0 and 3 are also marked off as shown, and point D projected from the base edge. From this point a line of indefinite length is drawn parallel to line J. In the end view a false length line is drawn between points on the top and base edges, determined by the position of the ordinates 3 and D. The joint curve—shown dotted—is cut by the false length at J. Point 3 is also marked 3' because 3' in the elevation lies directly behind it. Thus 3-D is the end view of the elevation line 3-D and it covers the edge line 3'-J. These lines are made into true lengths in the lay-out ; false length 3-D is measured along line D, and extended to 3, whilst 3'-J is marked off on line J and joined to 3'.

This lay-out can be used for the development of any job in which the end view has a less complex outline than that of the elevation.

FLUSH-SIDED BREECHES WITH PARALLEL ENDS

Fig. 19 depicts this method applied to the breeches piece shown in Fig. 18. Although an elevation is drawn, there is no necessity for this in actual practice, as the only dimension needed from it, the distance 0-P, can be taken direct from the working drawing. The job is tackled as follows : Draw an end view 0-6-G-A of one branch and on the top edge describe a semicircle. Divide this into six equal parts and draw in ordinates to the top edge, marking them 1, 2, 3, 4, and 5. Next describe the base semicircle, divide up as shown, and place in the ordinates B, C, D, E, and F. Draw in the shape of the joint face to the desired contour. Connect the points on 0-6 to the base edge points, joining 1 to B, 2 to C, and so on. Where these lines cut the joint curve gives points H, I, etc. Place in the remainder of the connecting lines to form the required triangles. The dotted lines are on the throat of the branch. Construct the lay-out in a suitable position by drawing an indefinite line and marking a point 0 and 6 on it. On each side of 0 and 6 measure off the lengths of the top edge ordinates, marking them with the appropriate numbers. Note that the points above 0 and 6 are marked with a dash to distinguish them from the points on the outer edge of the inlet. The edge line of the vertical plane is now placed in the lay-out by measuring the elevation distance 0-P from 0 and 6 and drawing a horizontal line from this point. Mark this line A, L, K, J, I, H, and G, as all these points lie on the vertical plane. Next measure down from P the ordinates B, C, and D from the base semicircle and draw lines parallel to line A . . . G, marking them with their respective letters.

To draft the pattern, pick up the false length 3-D and step it along D in the lay-out. Join to 3 on the perpendicular and draw the true length in a suitable position, using it as the middle line of the pattern. Next obtain the true length of 3-E,

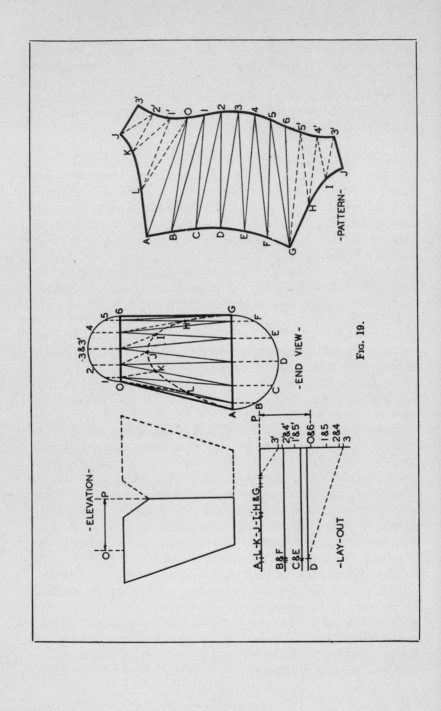

-PATTERN-

-END VIEW-

-ELEVATION-

A-L-K-J-I-H&G

B&F
C&E
D

3'
2&4'
1&5'
O&6
1&5
2&4
3

-LAY-OUT

Fig. 19.

along line E in the lay-out, by triangulating to 3. Describe
an arc with 3-E from 3 on the pattern middle line and cut it
in E with the spacing D-E. Continue constructing the pattern
triangles in this manner, but care must be taken after drawing
line 6-G that the preceding line 5-G be again picked up,
although on this occasion it is connected to 5' on the perpendicular because this line is on the throat, the top edge
points of which are marked with a dash in the lay-out. The
next line, 5'-H, is now marked off on the top line of the lay-
out and connected to 5'. Spacings G-H, H-I, etc., are taken
direct from the joint curve to the pattern. Complete the
pattern by working on the above lines, finally joining all
points with a smooth curve.

The branch when shaped up can be joined to the branch
described in Fig. 17 to form a breeches piece, which will
converge the air flow from two different points to a main pipe,
if such piping be fixed to a flat surface.

TWO-WAY TRANSITION ELBOW

The object shown in Fig. 20 is a transition elbow of original
design. It is used for conveying sawdust or similar material
from two adjacent hoods to a circular pipe, placed at right
angles to the hood outlets. The method of developing the
branches is similar to that of the previous example, with the
exception that, as the ends of the branches are elliptical in
shape to fit to the elbow joints, it is necessary to draw these
shapes on the joint lines A-G and 0-6. To do this, divide semicircles described on the elbow base lines into six equal parts,
place in the ordinates and produce the lines to the joint
lines as shown. From the points found, draw perpendiculars
of indefinite length and cut these off the lengths of the
ordinates from the elbow semicircles. Join the points with
fair curves, numbering the top semi-ellipse points 1, 2, 3, 4,
and 5, and the bottom semi-ellipse, B, C, D, E, and F. Triangulate the branch pattern as for Fig. 19, taking care to pick up
the spacings from the semi-ellipses for the pattern triangles
and not those on the elbow semicircles. Mark the second
branch off from the pattern and complete the laying-out of
the job by obtaining the elbow patterns, using the parallel-
line method of development.

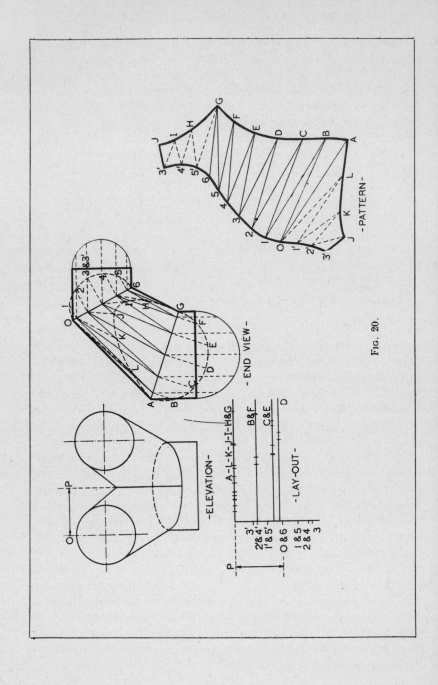

FIG. 20.

Chapter Six

OTHER SPECIAL TYPES OF JUNCTION PIECES

IN addition to the junction-piece problems already dealt with, there are other designs which call for some thought when tackling their patterns. Of these. the following examples are typical of many jobs encountered in everyday workshop practice. Fig. 21 shows the elevation and end view of a two-way junction piece with identical branches. Each

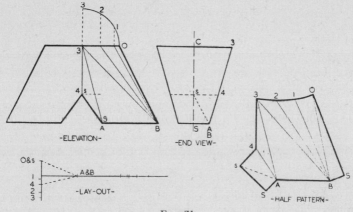

-ELEVATION- -END VIEW- -LAY-OUT- -HALF PATTERN-

FIG. 21.

inlet is rectangular, and from this shape the branches transform to a round main pipe. The end view of the job is symmetrical about its vertical centre line and C-3-4-s represents half of the shape of the joint face between the branches.

DEVELOPMENT OF RECTANGULAR TO ROUND TWO-WAY PIECE

To set-out the job, first draw the elevation 3-0-B-A-4 of one branch. On the top edge describe a quadrant, to represent a quarter section of the outlet, and divide it into three equal parts. Number the points 1, 2, and 3, and drop perpendiculars from them to the top edge. These lines—shown dotted—are ordinates required for the lay-out construction. Connect the points on the top edge to points A and B with false length lines,

49

to divide the elevation surface into triangles. Next draw the
end view of the branch. Project elevation point 4 across
horizontally to the end view. Distance *s*-4 is half the width
of the throat.

Now construct a lay-out in a suitable position. Draw a
vertical line and mark a point 0 and *s*. From this point
measure off and number the ordinates from the top edge.
Take distance *s*-4 from the end view and also measure it from
0 and *s*, marking point 4. Next take distance A-S from the end
view, measure from 0 and *s*, and from the point found draw a
horizontal line of indefinite length. Mark this line A and B.

To start the half-pattern draw a line *s*-S for the throat
seam in a convenient position. This line is marked 4*s* and
AS in the elevation, and it represents *three* lines, the seam
s-S, which is a true length ; the outside edge line 4-A, which
appears as a false length in the elevation ; and line *s*-A, shown
as a dotted line in the end view. Measure line 4*s*-AS along
line A and B in the lay-out, stepping it off from the corner of
the angle. From the point marked, join to points 4 and *s*
on the perpendicular to obtain true length lines.

Take the true length *s*-A and strike an arc from *s* on the
first pattern line previously drawn. Cut this arc in A with
distance AB-S from the end view, using S as centre. Next
take true length 4-A from the lay-out and describe an arc
from A in the pattern lay-out. Pick up distance *s*-4 from the
half-end view and cut the arc just drawn in 4, using *s* as
centre. For the next pattern triangle take false length 3-A
from the elevation, measure it along lay-out line A and B, and
extend from the point found to 3 on the perpendicular.
Describe an arc with this true length from A in the pattern
lay-out and cut this arc in 3, from centre 4, with distance 3-4
from the end view.

To continue the development, take elevation false length
3-B, measure it along the lay-out horizontal line, and triangulate
to 3 on the perpendicular. To add this line to the half-
pattern describe an arc from 3, and cut it in B from centre A,
with the elevation distance A-B. Next take false length 2-B
(the full line) and step off on the lay-out horizontal line.
Extend from the point to 2 on the perpendicular, and with
this distance strike an arc from B in the development. Cut

this arc in 2 from centre 3, with the spacing 2-3 from the quadrant on the elevation top edge.

Similarly, find the true lengths of 1-B and 0-B, and add to the half-pattern. Finally, take the edge line 0-B, using it this time as a true length, and describe an arc from 0. Take AB-S from the end view and cut the arc just drawn in s, from centre B. Join points 0 to 3 with an even curve. The full pattern can be obtained, if desired, by reproducing the lines already drawn.

THREE-WAY JUNCTION PIECE

The next example, illustrated in Fig. 22, is a three-way junction piece, the inlet and outlet centres of which are on

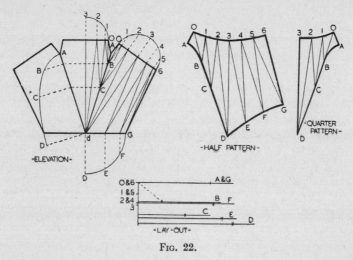

Fig. 22.

the same vertical plane. The outside branches are identical in shape and size, thus only the development of a half-pattern is necessary. A quarter-pattern for the middle branch is shown, it being used as a template to obtain the full pattern. Both developments are obtained using the same lay-out. The half-pattern for the outside branches is drafted as follows :—

Draw the elevation as shown and describe a semicircle on the top edge of one of the outside branches. Divide it up into six equal parts and draw perpendiculars to the top edge. Describe quadrants on the top edge of the middle branch and the base edge of the junction piece. Divide these quadrants up into three equal parts and draw in ordinates. Number and letter all the points as shown.

Next draw ordinate d-D equal to the base radius on the joint line d-A, and describe an elliptical curve from A to D. This is the half-section of the joint between the branches. Divide the curve into three suitable parts, A-B, B-C, and C-D, and draw ordinates from B and C to the joint line. Project the points horizontally to the opposite joint line, and connect the edge points with false length lines to form the elevation triangles.

Draw the perpendicular of a lay-out and from a point 0 and 6 draw an indefinite horizontal line. Measure the top edge ordinates from 0 and 6 and mark the points with their appropriate numbers. Now measure from the point the joint and base line ordinates, draw indefinite horizontal lines from the points, and letter as shown.

To commence the drafting of the half-pattern, draw the seam, which is the true edge line 0-A, in a suitable position. Pick up false length 0-B on the outside branch, mark it off on lay-out line B, and extend to 0 on the perpendicular. Strike an arc with this distance from 0 on the pattern seam and cut it in B, with spacing A-B from the quarter-ellipse, using A as centre. Next take elevation length 1-B, step off on lay-out line B, triangulate to 1 on the perpendicular, and swing an arc with the true length from B in the pattern lay-out. Cut this arc from centre 0, with spacing 0-1 from the inlet semicircle.

Take false length 1-C, mark it off on lay-out line C, extend to 1 on the perpendicular, and with this distance describe an arc from 1 in the development. Strike an arc from B with spacing B-C from the quarter-ellipse, to cut the arc previously drawn in C. Continue this procedure with false length 2-C and the remaining connecting lines on the elevation surface. Each false length in turn is extended to a true length by measuring it along its correctly lettered lay-out line and joining to the appropriate number on the perpendicular. The pattern triangles are built up with the true length lines and the spacings from the half-sections on the elevation edge lines. Development of the quarter-pattern for the middle branch is accomplished in exactly the same manner and should present no difficulties if the example given is carefully followed.

ALTERNATIVE DESIGN FOR THREE-WAY PIECE

Instead of being formed of three separate branches, the job can be made in two parts. This method, illustrated in Fig. 23, has certain advantages, one of which is that

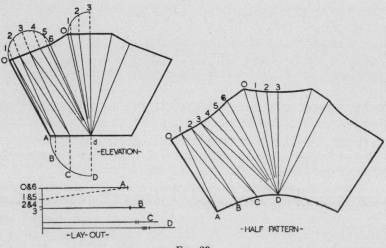

FIG. 23.

troublesome fitting of the branch joints is avoided. The patterns, however, require a certain amount of care in shaping up.

Between the branches of the way-piece are formed four triangular connecting pieces, one of which is marked 6-d-0 in the elevation. The top edges 0-6 are the joints between the front and back triangles. No difficulty should be experienced in drafting the half-pattern, as both this and the lay-out construction are similar in principle to previous examples. Use the half-pattern as a template for the other side of the job. Note that all the connecting lines, with the exception of the seams, marked 0-A, are false lengths, and, as such, must be made into true lengths in the lay-out. The top edge joint lines 0-6 are true lengths in the elevation. Care must be taken to arrange the elevation triangles as shown to enable the patterns to be shaped up in the folders with the minimum of trouble.

TWO-WAY PIECE WITH UNEQUAL BRANCHES

Fig. 24 shows a junction piece with inlets of different diameters which transform to a rectangular base, both branches being joined up with triangular pieces marked in the elevation 6-B-0'. The lay-out for this job has the ordinates from the inlet half-sections marked off on the perpendicular from the same point, as shown. From this point is also marked off the base half-width a-A from the half-end view and a horizontal line of indefinite length is drawn through the point. On the left-hand side of the lay-out is measured off the false lengths from the smaller branch, the other branch connecting lines being stepped off on the right-hand side. Note that the edge line 6-0' is a true length and that the edge lines 0-Aa and 6'-Cc represent both true edge and false length lines ; all the remaining connecting lines between the inlet and base edges must be made into true lengths in the lay-out.

Start the half-pattern by drawing edge line 0-a in a convenient position. Next measure this distance off from the foot of the perpendicular on base line A and B in the lay-out and join to 0 and 6. With this true length describe an arc from 0 on the pattern edge line, and cut in A from centre a with half-width A-a from the half-end view. Take elevation length 1-A, measure it along lay-out line A and B, and connect to 1 on the perpendicular. Describe an arc in the pattern with this true length from centre A and cut it in 1 with spacing 0-1 from the top semicircle.

Continue the development on similar lines up to and including pattern line 6-B, then take elevation length 0'-B, measure it off on lay-out line B and C from the foot of the perpendicular, and join to 0' and 6'. From B in the pattern draw an arc with the true length found, and cut in 0' from centre 6 with distance 6-0' from the elevation. Complete the half-pattern in the usual manner as shown in the diagram.

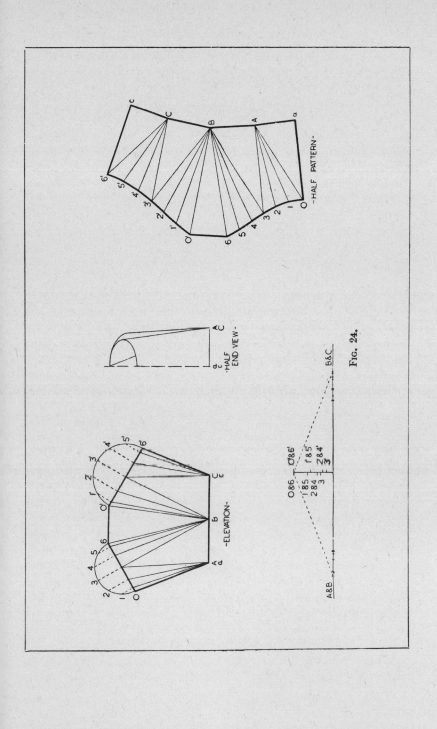

FIG. 24.

Chapter Seven

CONES INTERSECTING CYLINDERS

AIR-DUCT branch connections are often made conical in shape and of long taper. Thus the drafting of their patterns by the radial-line method is sometimes not a practical proposition, although such a course is usually

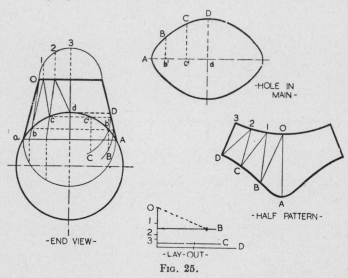

FIG. 25.

advocated for work of this nature. Fig. 25 shows the end view of a conical branch connection to a round main. The centre line of the branch is perpendicular to that of the pipe. By using the lay-out system it is possible to obtain the pattern by triangulation, after the shape of the pipe hole in the flat has been found. It is necessary to find this shape, because the hole perimeter gives the correct length of the joint line between the parts.

CONICAL BRANCH CONNECTION

The setting-out of the job to obtain the pattern and hole shape is as follows : First draw the end view of the duct,

56

as shown, and describe a semicircle on the top edge. Divide one-half into three equal parts and number the points 0, 1, 2, and 3. Draw perpendiculars from the points to the top edge to obtain ordinates for the lay-out construction. Next mark the joint points *a* and A, and join up to form the cone base. Describe a semicircle on *a*-A, and draw perpendiculars to the base line from three equally spaced points on the semicircle, as shown. Connect the top and base edge points to cut the cylinder circumference at *b*, *c*, and *d*. Join 0 to *b*, 1 to *c*, and 2 to *d*, to form triangles. Draw horizontal lines—shown dotted —from *b*, *c*, and *d* to obtain points *b'*, *c'*, and D. The distance *d*-D is the cone half-width at the point where it meets the cylinder. Describe an arc from the edge line, on line *c*, using as radius the distance between the edge line and the vertical centre line. Drop a perpendicular from *c'* to cut the arc in C. Distance *c'*-C is the hole half-width at *c'*. Repeat this construction for *b'*-B.

Now set-out the shape of the hole in the cylinder. Draw an indefinite line in a suitable position and mark a point A. From A step off distances A-*b'*, *b'*-*c'*, and *c'*-*d* from the end view.

Erect perpendiculars from the points and cut them off in length the distances *b'*-B, *c'*-C, and *d*-D. Draw a smooth curve through the points to complete one-quarter of the hole. The full hole shape requires the repetition of the above construction.

To construct the lay-out draw an indefinite line for the perpendicular and mark a point, 0. Measure from 0 the ordinates on the top edge of the connection, and mark the points 1, 2, and 3 respectively. Next measure from 0 the distances *b'*-B, *c'*-C, and *d*-D. From the points found draw indefinite horizontal lines and mark them B, C, and D, as shown.

To start drafting the pattern draw a line equal in length to the edge line 0-*a*. Take false length 0-*b* and mark it off on line B in the lay-out, measuring from the corner. Open the dividers from the point marked to 0 on the perpendicular, and with this true length strike an arc from 0 on the pattern line 0-A. Cut this arc in B, using A as centre, with the spacing A-B from the hole perimeter. Take 1-*b* from the end view,

measure it along lay-out line B and join to 1 on the perpendicular. Describe an arc with this distance from B in the
pattern lay-out and cut it in 1 from 0, with the spacing 0-1
from the top semicircle. Next pick up 1-c, measure off on
lay-out line C, and triangulate to 1 on the perpendicular.
With this true length swing an arc from 1 in the pattern lay-
out, and using B as centre, cut it in C with spacing B-C. Repeat
this construction for the remainder of the triangles. Step
each false length in turn on its appropriate lay-out line, triangulate to the correct perpendicular number, and build up
the pattern triangles in the usual manner. Join all the points
with an even curve, the half or full pattern being drafted as
desired.

ALTERNATIVE METHOD FOR PATTERN

Some of the geometrical construction described can be
simplified if desired, for workshop use. Although branch connections or " shoes " are often conical in shape, it is not
absolutely necessary that they be made portions of right or
oblique cones, as an approximate shape is suitable in many
cases. By constructing a hole of convenient shape and size
it is possible to triangulate the connection surface with a
minimum of lines. Where the " shoe " is of irregular shape
this is the most practical method to adopt, although it must
be clearly understood that the shape of the job depends upon
the contour of the joint line or hole perimeter.

Fig. 26 shows a view of a branch connection similar to the
last example. Draw a half-end view in a convenient position,
as this is all that is necessary for the development of the
pattern. Divide the joint line into three suitable parts and
mark the points A, b, c, and d. On the top edge describe a
quadrant, divide it into three equal parts, and number the
points 0, 1, 2, and 3. Draw in the ordinates and connect the
points on the top edge to points b, c, and d on the joint line
by means of false length lines. Project d to the edge line,
obtaining D. This distance d-D is the half-width of the hole.

For the hole in the main draw an indefinite horizontal
line and mark off on it the spacings A-b, b-c, and c-d from the
joint line. Draw a perpendicular through d and mark off
d-D each side of the horizontal line. This distance is the
minor axis of an ellipse of which the major axis is twice the

length of A-*d*. An elliptical curve drawn through the points is the required shape of the hole. Finally, erect perpendiculars from *b* and *c* to cut the perimeter at B and C.

Draw a lay-out in a suitable position. Measure off and number ordinates 1, 2, and 3 from the top edge of the job

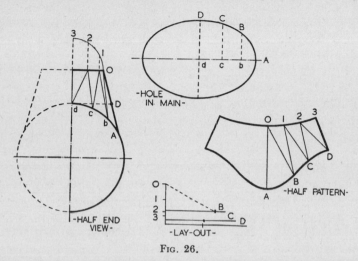

FIG. 26.

from a point marked 0 on the perpendicular. Next measure from 0 distance *d*-D from the hole, and draw a horizontal line D from the point. Take *b*-B and *c*-C from the hole and also mark off from 0. Draw lines parallel to D from the points and mark them B and C respectively. The half-pattern is developed in exactly the same manner as for the previous example.

CONICAL HOPPER ON INCLINED PIPE

In Fig. 27 is depicted a conical hopper fitting at an angle to a round pipe. When the job is in position the top of the hopper lies parallel to the ground, but the pattern is most easily tackled from the views drawn as shown. In order to obtain the correct shape of the joint line between the hopper and the pipe it is necessary to draw an elevation and a half-end view. First draw the elevation and describe semicircles on the top and base lines of the cone frustum. Divide each of them into six equal parts and draw perpendiculars—shown dotted—to the edge lines. Now join up the edge points with straight lines. Next construct a half-end view projected from the elevation, with the semi-ellipses drawn in, as shown. Join the top points b, c, d, e, and f to the base points B, C, D, E, and F, intersecting the semicircle at points $1'$, $2'$, $3'$, $4'$, and $5'$. Draw dotted lines from these points parallel to the cylinder centre line, to cut similar lines in the elevation. Through the points draw the joint line, and then join them up to the edge points on A-G to form triangles.

Next draw a perpendicular from 3 on the joint line to 0-6, marking X. Construct the shape of the hole in a convenient position. Draw a line equal in length to 0-6, mark off the distance 6-X and erect a perpendicular at X. Measure off from X the distances $0'$-$1'$, $0'$-$2'$, etc., from the half-end view. Draw horizontal lines from the points as shown, and make them equal in length to those distances which lie between similar numbered points on the joint line and 3-X in the elevation. Mark the points on the hole perimeter, 1, 2, etc., and join them up with a smooth curve.

To construct the lay-out draw a vertical line and mark a point 0 and 6. Measure the horizontal distances between a-G and points $1'$, $2'$, $3'$, etc., in the half-end view and mark them off on the lay-out perpendicular. Number the points 1, 2, 3, 4, and 5 respectively. Next draw a horizontal line from 0 and 6 and mark it A and G. Take the distances from between A-G and the semicircle in the elevation, and measure off from 0 and 6. Draw horizontal lines from the points and mark them B and F, C and E and D respectively.

For the pattern, first draw the edge line A-0. Next take 0-B from the elevation, measure it along lay-out line B, and triangulate to 0 on the perpendicular. Strike an arc with this true length from 0 on the edge line and cut it in B with the spacing A-B from the cone. The rest of the half-pattern is developed on similar lines and should present no difficulties if the example given is closely followed. Note that spacings 0-1, 1-2, etc., are taken from the hole perimeter.

AN EASIER METHOD OF SETTING-OUT

The hopper pattern can be drafted with fewer lines by drawing an approximate elevation joint line and triangulating

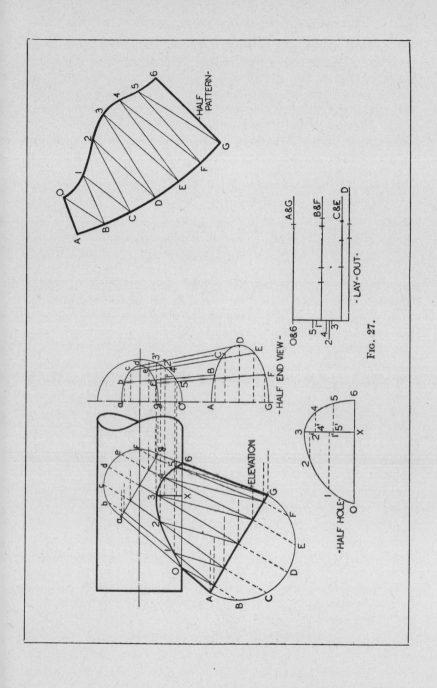

-HALF PATTERN-

-HALF END VIEW-

-ELEVATION-

-HALF HOLE-

-LAY-OUT-

O&6

B&F

C&E

A&G

D

Fig. 27.

the surface of the job to the contour of the pipe hole. This method, which is illustrated in Fig. 28, can be used for many similar jobs in the workshop. First draw the elevation as shown, and describe a semicircle on A-G. Divide it up into six equal parts and draw perpendiculars back to the edge line. Drop a perpendicular from a suitable point x' on the pipe centre line and project X on the edge line across to the perpendicular marking X'. Draw a perpendicular from the point and make it equal to X-D from the semicircle. Next measure x'-d' on the pipe centre line, equal to x-d, and join d' to D'. Describe an arc from x' with radius x'-0' to cut d'-D' in 3'.

Divide the distance 0'-3' into a suitable number of parts and from the points draw dotted lines of indefinite length parallel to the pipe centre line. At the point where the dotted line from 3' cuts the hopper centre line, mark 3. Draw a suitable curve from 0 to 6, through 3, to cut the dotted lines at 1, 2, 4, and 5. Join these points by means of false length lines to the edge points on A and G.

Now construct the shape of the half-hole. On a vertical line mark off spacings 0'-1', 1'-2', and 2'-3', and draw horizontal lines from the points. Measure off from the perpendicular points 0, 1, 2, etc., using the same distances as between the joint line points and the perpendicular x'-X'. Join up the points on the hole perimeter with a smooth curve. For the lay-out, make the distances from 0 and 6 on the perpendicular equal to the ordinates between x'-0' and points 1', 2', and 3' on the arc. Number as shown. Also measure from 0 and 6 the appropriate distances from the base edge semicircle, and mark horizontal lines drawn from the points, as shown.

Develop the pattern in the usual manner by obtaining the true lengths in the lay-out, using the spacings 0-1, 1-2, etc., from the hole perimeter for the top edge points of the triangles. Although the hopper will not come out as a true portion of a cone, it will be practically the same in appearance ; in fact, if the approximate joint line is drawn by a workman with an experienced eye there will be no apparent difference at all. The fit between the hopper and pipe, as in the last problem, should be perfect if the patterns are carefully drawn.

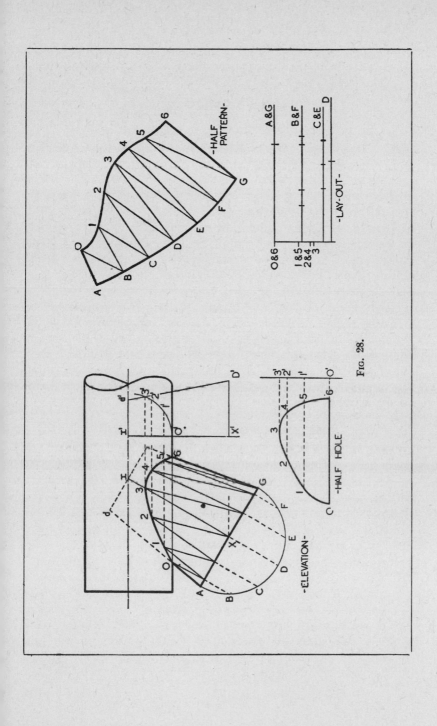

FIG. 28.

Chapter Eight

CONICAL INTERSECTIONS

PATTERNS for jobs which are made up of cone frustums fitting together are rather difficult to develop because of the necessity of locating the joint or intersection curve between the parts. Further complications arise if the frustums are of long taper, as this usually means that the patterns must be triangulated. Simplification of the problems involved can be accomplished by using the elevation surfaces as a basis for obtaining the pattern triangles, as this eliminates much tedious projection work required when triangulating from a plan of the job.

The shape of the line of intersection between the cone frustums is best obtained using the cutting plane method in which section lines are drawn across the elevation of the job in suitable positions. These lines are the elevations of sections made by cutting planes, and on them are located the joint line points derived from the intersections of the parts as depicted in the sections.

USE OF CUTTING PLANE METHOD

In Fig. 29 is shown the elevation of a conical branch connection to a tapered main connecting piece. A section line X-Z is drawn across the elevation. The section of the horizontal cone at the cutting plane gives the circle z-Z; the slanting cone section appears as an ellipse, of which x-X is the major axis. For the minor axis a section line W-Y is drawn through c, in the elevation, square to the slanting cone centre line, and this cutting plane presents the circle W-Y. Point c, which is the centre of x-X, is projected to this section to obtain distance c-c'. By this means the half-width of the smaller cone at point c is found. At the intersection of the circle and ellipse in the vertical section on X-Z is afforded a point C, which is projected horizontally to the section line X-Z in the elevation to determine the similar lettered point on the line of intersection. Further points on this line can be found, using the same principle, of which a simplified construction is shown in Fig. 30.

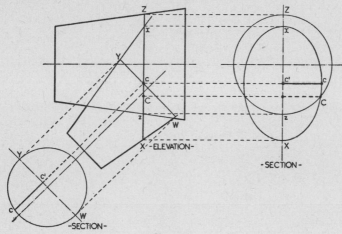

FIG. 29.

Using this method of cutting planes also enables the lay-out triangulation system to be used for the drafting of the patterns, as part of the construction can be adapted for this purpose. The dotted line drawn from C to X-Z in the vertical section is an ordinate used in the lay-out construction, and the arc z-C is the distance required for the half-hole in the tapered piece at this point.

SETTING-OUT FOR CONICAL PIPE CONNECTIONS

Draw the elevation as depicted in Fig. 30. Produce the edge line 0-A of the branch indefinitely, and draw vertical section lines to intersect it through suitably spaced points B′, C′, and D′ on the edge line between A and E. To obtain point D on the joint curve, bisect section line f-g to find point d. Through d draw m-p perpendicular to the branch centre line. Describe an arc from m with radius m-p and meet this in t with a line from d, perpendicular to m-p.

The distance d-t is half the minor axis of the elliptical section of the cone at the cutting plane. Project d horizontally to the base edge of the horizontal frustum and draw a perpendicular to the edge line from the point found. Make

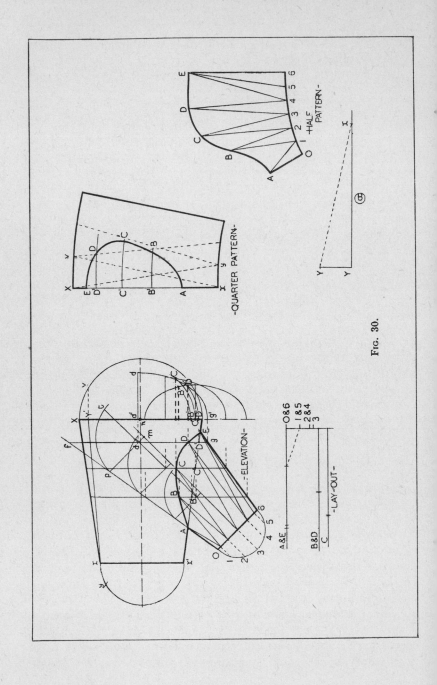

-HALF PATTERN-

-QUARTER PATTERN-

-ELEVATION-

-LAY-OUT-

A&E O&6
      1&5
      2&4
B&D   3
C

Fig. 30.

d-d' equal in length to *d-t*. Now project *g* on the section line to *g'* and draw an elliptical curve between *d* and *g'* as shown. Next project D' to the base edge *g'*-X, and with radius *n*-D' describe an arc, with *n* as centre, to cut the quarter-ellipse in D. Project this point back to section line *f-g*, marking the point D. The dotted line between the base edge and D is an ordinate required later on in the lay-out construction.

Repeat this setting-out for the other section lines to obtain points B and C on the joint line, and draw in the curve between the points. (The half-section of the small cone on the section line through B' is drawn as a semi-ellipse to illustrate that point B is on the top half of this curve.) Draw a semicircle on edge line 0-6, divide it into six equal parts, and project the points back to the edge line. Join the points on 0-6 to A, B, C, D, and E on the joint line by means of false length lines to obtain the elevation triangles, as shown.

PATTERN FOR THE MAIN CONNECTING PIECE

The development for the horizontal tapered piece pattern is as follows, one-quarter only being shown. First describe a semicircle on the base edge in the elevation, using *n* as centre. Make the distance X-Y one-sixth part of the semi-circle and draw ordinate Y-Y', shown dotted. Draw on the top edge a semicircle and make *x-y* a one-sixth part of it. Join *x* to Y' with a false length line.

Next draw a right angle in a convenient position, as at (*a*), and mark off ordinate Y-Y' on the perpendicular and false length *x*-Y' on the base line. Join *x* to Y to obtain a true length for the pattern. To start the pattern, draw the edge line *x*-X in a suitable position. Take the true length *x*-Y and describe with it an arc from *x* on the edge line. Cut this arc in Y with spacing X-Y from the base semicircle. Next, with true length *x*-Y describe an arc from X and cut it in *y* with spacing *x-v*. Repeat this construction for the remainder of the development and join the points with an even curve.

From *x* on the edge line measure off distances *x'*-A, A-B', B'-C', etc., from the elevation and draw arcs of indefinite length from the points. These arcs are best obtained by measuring them off equidistant to the curve *x-y*. Cut the arcs off equal in length to arcs B-B', C-C', and D-D' on the

base edge of the horizontal frustum and join up the points
with an even curve, thus obtaining the shape of half the hole
in the tapered piece.

Now construct a lay-out for obtaining the true lengths of
the elevation triangles. Draw an indefinite vertical line and
mark a point 0 and 6. From this point draw a horizontal line
of indefinite length and mark it A and E. Take ordinates
B, C, and D from the edge line g'-X and mark them down
the lay-out perpendicular from 0 and 6. The ordinates are
shown dotted, and in this example ordinates B and D are the
same length. Draw horizontal lines from the points on the
perpendicular and mark them as shown. Next, from 0 and 6
measure off the distances, shown dotted, between the semicircle and edge line 0-6 on the slanting cone. Mark the points
with their respective numbers.

To start the branch half-pattern take edge line 0-A from
the elevation and draw in a convenient position. Pick up
false length 1-A and measure it off from 0 and 6 on line A
in the lay-out. Extend the dividers from the point marked
to 1 on the perpendicular, and with this true length strike an
arc from A on the first pattern line. Cut this arc in 1 with
spacing 0-1 from the branch semicircle, using 0 as centre.
Next take false length 1-B, step it from the corner of lay-out
line B, and triangulate from the point found to 1 on the perpendicular. Describe an arc with this distance from 1 in the
pattern lay-out and cut it in B with spacing A-B from centre A.
This spacing A-B is taken from the half-hole perimeter in
the quarter-pattern of the tapered piece.

Now take false length 2-B, measure it along B line in the
lay-out and extend to 2 on the perpendicular. Describe
an arc from B in the pattern lay-out with this true length and
cut it in 2 with spacing 1-2, using 1 as centre. The remainder
of the development is on similar lines, each false length in
turn being extended to its true length in the lay-out and used
to form the pattern triangles. Take care to place each false
length on its similar lettered line in the lay-out and to triangulate to its appropriate number. The top-edge spacings
of the half-pattern are all taken from the hole perimeter, and
if carefully drawn there should be no trouble in completing
the patterns to give a good fit between the parts when made up.

AN EASIER METHOD OF SETTING-OUT

A simpler method of obtaining the branch connection is shown in Fig. 31. One cutting plane only is used to find a point C on the line of intersection. A suitable curve is drawn through C from A to E, and from this is obtained measurements for the shape of the hole in the horizontal tapered piece. It must be emphasised that this is a method for workshop use, as the branch connection will not, strictly speaking, be a portion of a right cone, but it is correct enough for many practical purposes. As the shape of the branch is governed by that of the hole in the tapered piece, the fit between the two pieces of duct should leave nothing to be desired.

Lay-out the job in the following manner : First draw the elevation as shown. Divide distance A-E on the edge line of the tapered pipe into a suitable number of parts at points b, c, and d. Erect perpendiculars to the horizontal centre line from b and d. Draw a vertical section line through c from p to meet in t, edge line 0-A produced. Bisect p-t in m. Through m draw a line perpendicular to the branch centre line, marking o. Describe an arc from o with radius o-P. Raise a perpendicular from m to meet the arc in h. This distance h-m is the half-width of the cone at point m.

Now draw m-n, which is equal to h-m, square to the section line and connect points n and p with an elliptical curve. Next, using c' on the section line as centre, draw an arc with radius c-c' to cut the elliptical curve in C′. Project C′ horizontally back to the section line to obtain C. Draw in the joint curve through this point from A to E to cut perpendiculars b-b' and d-d' in B and D respectively. Describe arcs from b and c, using b' and c' as centres. Draw dotted lines to cut these arcs, parallel to the horizontal centre line, from B and D to obtain ordinates.

Draft the quarter-pattern for the tapered piece in similar fashion to the last example. Note, however, that the true length x-Y in the elevation is obtained by erecting a perpendicular to x-Y′ from Y′, equal in length to ordinate Y-Y′. Points A, b, c, d, and E are measured off on the edge line of the quarter-pattern similar to the points on the elevation edge line x'-E. Arcs b-B, c-C, and d-D are the same length as the

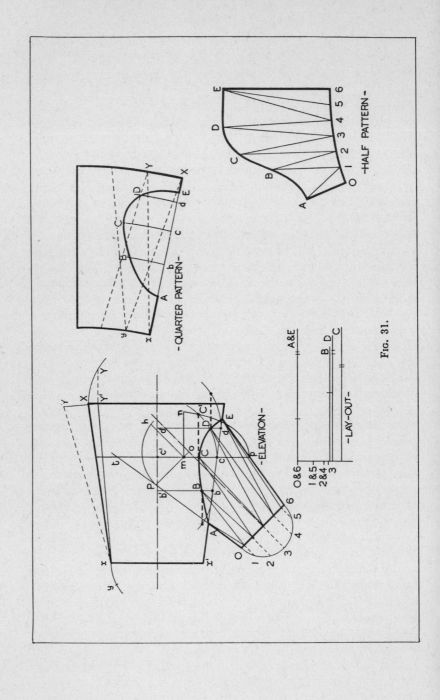

FIG. 31.

arcs from b, c, and d in the elevation, c-C, for instance, being equal to c-C′.

For the lay-out, ordinates B, C, and D—the heavy dotted lines—are measured off from 0 and 6 on the perpendicular and indefinite lines drawn from the points. The ordinates from the small end of the branch are also marked off from the same point and numbered as shown.

PATTERN FOR CONICAL BRANCH

First take the branch edge line 0-A from the elevation and draw it in a convenient position as the first pattern line. Next take false length 1-A, measure it along the top line of the lay-out, and from the point found join up to 1 on the perpendicular. With this true length describe an arc from point A on the pattern line and cut it in 1 with spacing 0-1 from the branch semicircle, using 0 as centre. Now pick up elevation length 1-B, measure along lay-out line B, and join to 1 on the perpendicular. Describe an arc with the true length from point 1 in the pattern lay-out, and cut in B with spacing A-B from the hole perimeter in the main connecting piece quarter-pattern.

Next take false length 2-B from the elevation and measure it along lay-out line B. From the point marked, join to 2 on the perpendicular. Strike an arc with the true length, from point B in the pattern, and cut in 2 with spacing 1-2 from the branch semicircle. Take elevation length 2-C, mark off along lay-out line C, and connect to 2 on the perpendicular from the point found. With the true length describe an arc from 2 in the pattern, and cut it in C from point B with spacing B-C from the hole perimeter in the main quarter-pattern.

Continue the development on similar lines, as shown in the diagram. After marking-off pattern line 6-E the full pattern can be drafted by reproducing in reverse order the lines already drawn.

There are many variations of conical intersection work, but the laying-out of the patterns should present little difficulty if the principle underlying the use of the cutting-plane method is clearly understood.

Chapter Nine

TWISTED AND OFF-CENTRE TRANSFORMERS

AS explained in the first chapter on methods of pattern development, one of the difficulties of laying-out work from blue prints is that certain views which are considered necessary for pattern-drafting purposes are not given by the draughtsman. Much trouble is caused when tackling the patterns for off-centre and twisted air-duct transformers through the necessity of projecting additional plans and elevations from the given views of the plant lay-out. With the lay-out system, however, the views shown on the drawing can be used as a basis for obtaining the pattern true lengths. The whole procedure is thus very much simplified and valuable time saved.

FAN OUTLET CONNECTING PIECE

Fig. 32 shows a working drawing (minus the dimensions) of a connecting piece which transforms from the rectangular outlet of a fan to a round discharge pipe. This pipe lies at an angle to both the horizontal and vertical planes. No other view of the job is shown on the drawing, but if a plan were projected from the elevation it would show the fan outlet and pipe twisted to each other. This plan view is not necessary for development purposes as the elevation or end view of the job is used.

In the drawing the centre lines of the duct are shown inclined to the base line in each view ; the angles are marked A and B respectively. These centre lines are the projected or " false " lengths of the true centre line of the duct. If desired, the true or resultant angle at which the pipe lies to the horizontal plane can be calculated by the following trigonometrical formula in which A is the elevation angle, B the end view angle, and θ the resultant angle.

$$\cot \theta = \sqrt{\cot^2 A + \cot^2 B}.$$

EXAMPLE.—Given that $A = 60°$, $B = 70°$,

then
$$\cot \theta = \sqrt{\cot^2 A + \cot^2 B}$$
$$= \sqrt{\cdot 5774^2 + \cdot 3640^2}$$
$$= \sqrt{\cdot 465886 \ldots}$$
$$= \cdot 682 \ldots$$
$$= 56° \text{ nearly.}$$

It is not essential to know this angle for this particular pattern, but the formula has been included because it can be

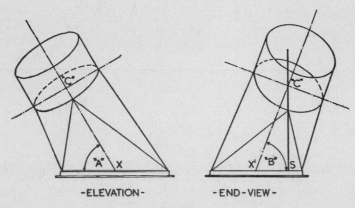

-ELEVATION-   - END-VIEW-

FIG. 32.

used to obtain the true angle for any kind of compound-angle duct work.

Before the pattern can be drafted it is necessary to draw an elevation showing a view of the transformer face on the vertical plane. In the working drawing the centre lines of the top are inclined to the base in both views, but if an elevation were drawn of the job with the top centre line parallel to the horizontal plane the elevation centre line " C "-X would then be perpendicular to that plane. Fig. 33 shows a perspective view of the transformer placed in this position with a vertical plane cutting through the centre of the top face, parallel to the base edge A-B. The plane is placed in this way to illustrate a method by which a view of a section through the transformer is obtained. This section is projected on to a vertical plane placed adjacent to the object, the lengths of the projectors being used to determine points in the lay-out construction.

On the cutting plane is shown the elevation centre line " C "-X. A line drawn on the top face from the edge point 3 to the centre " C " is inclined to the elevation centre line at an angle determined by the length of the centre line of the transformer. This true length is actually the hypotenuse of a right-angled triangle—shown shaded—of which " C "-X is the perpendicular and the end view offset distance the

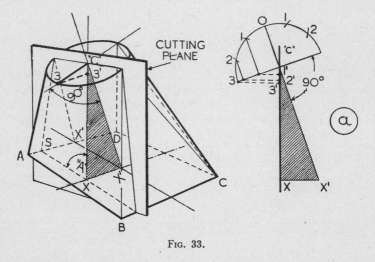

Fig. 33.

base. Point 3 on the top edge is projected to 3′ on the plane. " C "-3′ is the length of half the minor axis of an ellipse, of which the top diameter forms the major axis. This ellipse is the required view of the transformer face.

To obtain points on the ellipse for the elevation outline, first draw, in a suitable position, a vertical line, as at Fig. 33 (a), and mark off on it " C "-X from the elevation on the working drawing. This line also represents the cutting plane edge. From X draw a horizontal line and make it equal to S-X′ from the end view. Draw a line from X′ through " C." Also through " C " draw an indefinite line at right angles to the hypotenuse and describe a semicircle on it, using " C " as centre, with a distance equal to the top face radius. Divide the semicircle into six equal parts and mark the points 0, 1, 2, and 3 as shown. Draw perpendiculars to the top edge line and project lines—shown dotted—horizontally to the cutting plane edge, to obtain points 1', 2', and 3'.

The ellipse is constructed on the top line 0-6 of an elevation drawn as shown in Fig. 34. This is a correct orthographic projection of the job on the vertical plane, showing the ellipse in its right relation to the true length A-B, but the ellipse can be drawn, if desired, on the top centre line of the elevation in Fig. 32, as the pattern can be correctly triangulated *by the lay-out system*, using this profile. The construction of a lay-out to obtain the true lengths of the elevation lines is very simple, as the positions in it of the vertical planes A-B and C-D in their relation to the cutting plane are determined by those distances on the end view base edge which lie each side of this plane. In actual practice the end view offset distance S-X' (Fig. 33) is added to half the base width, obtaining D-S, and then S-X' is deducted from the half-width, giving A-S. These two distances are marked off on the lay-out perpendicular from a point denoting the centre line 0-6 of the top face.

OBTAINING THE TRANSFORMER PATTERN

The construction required to obtain the pattern is as follows : First draw an elevation A-0-6-B, as shown in Fig. 34. Describe a semicircle on 0-6 and divide it into six equal parts, marking the points 1, 2, 3, etc. From the points draw lines of indefinite length through 0-6 and at right angles to this line. Next take the distance " C "-3' from Fig. 33 (a) and mark it off each side of 0-6, using " C " as centre, on line 3. Similarly, take " C "-2' and mark it off on lines 2 and 4. Repeat this construction with " C "-1' on 1 and 5. Join all the points with a smooth curve to complete the ellipse. Next divide the elevation surface into triangles. Join the ellipse points to corners A and B as shown, drawing full lines for the front and dotted lines for the back of the transformer.

Now construct the lay-out in a convenient position. Draw a line for the common perpendicular of the triangles, mark a point 0 and 6 and measure from it the projectors from

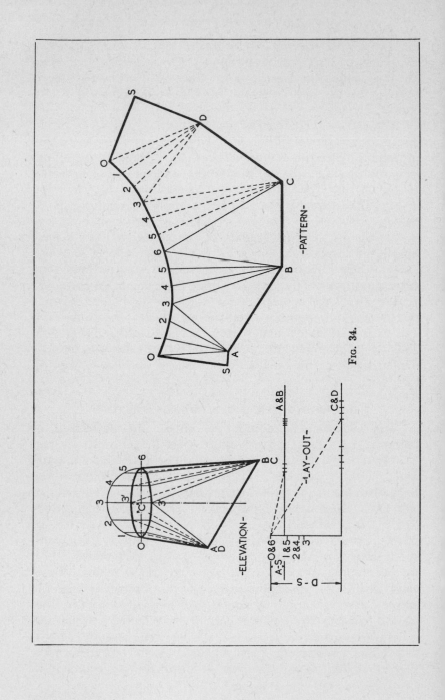

FIG. 34.

1-1', 2-2', and 3-3' in Fig. 33 (a). These are the dotted lines between the inclined top line and the cutting plane edge. Number the points on the perpendicular 1 and 5, 2 and 4, and 3. Measure from 0 and 6 the base distances A-S and D-S. Draw horizontal lines from the points found and mark the lines A and B and C and D respectively.

To start the pattern, take the edge line 0-A from the elevation and draw it in a suitable position. As this is the seam, mark the top point 0 and the base point S. Next step 0-A along line A and B in the lay-out, measuring from the corner of the angle. Take the true length between the point marked and 0 on the perpendicular and describe an arc from 0 on the seam. Using S as centre, cut this arc in A with the base distance A-S from the end view or lay-out perpendicular. Take the false length 1-A, that is, the full line between A and the nearest point on the ellipse in the elevation, and measure it along lay-out line A and B. Join to 1 on the perpendicular and describe an arc with the true length from A in the pattern lay-out. Cut this arc in 1 with the spacing 0-1 from the elevation semicircle, using 0 as centre. Measure the false length 2-A along A and B in the lay-out and triangulate to 2 on the perpendicular. Add this true length to the pattern lay-out in similar fashion to the other true lengths, cutting it with spacing 1-2.

Continue the development on these lines for the front triangles of the job. The elevation edge line 6-B has its true length placed in the pattern to form one side of the end triangle B-6-C. To obtain the true length of 6-C, measure the edge line along line C and D in the lay-out and join to 6 on the perpendicular. Describe an arc from 6 in the pattern lay-out and cut it in C with the base width B-C, using B as centre. Build up the back triangles from the false lengths shown dotted in the elevation. Step each line in turn along C and D in the lay-out and triangulate to the appropriate perpendicular number. Place them in the pattern in the same manner as for the first triangles. The last triangle, D-0-S, has D-S made equal in length to D-S in the lay-out. Finally, join all the points on the top edge of the pattern with a smooth curve and the base points with straight lines.

As an alternative method the pattern can be triangulated

from an end view of the transformer, with a top edge inclined
to the base at an angle similar to that of the top edge at (a) in
Fig. 33, but this calls for a slightly more complicated type of
lay-out, so the elevation method is the more suitable for most
jobs. Working from an elevation in conjunction with a lay-
out constructed on the cutting plane principle makes possible
the easy solution of many twisted transformer and transition
piece problems once the system is clearly understood.

TWISTED TRANSFORMER WITH TOP INCLINED TO A SQUARE BASE

This duct, shown in Fig. 35, is similar to the previous
problem with the exception that the transformer top and
base are situated about the vertical centre line. The job twists
from an inclined round pipe to a square base. In the elevation
and end views the top centre lines are inclined to the horizontal
plane, being at right angles to the centre line of the pipe.
The vertical centre line in the end view can be considered as
the edge of a cutting plane ; the view of the inclined top face
on this plane is found by the construction shown at Fig.
35 (a). In this view " C "-X is taken from the elevation and
X-X' from the end view ; this is the offset distance S-X'.
At right angles to the hypotenuse is drawn an indefinite line
through " C." From " C " are marked off ordinates taken
from a semicircle drawn on the elevation line 0-6, giving
points 1, 2, and 3. Projectors from these points are drawn to
" C "-X, giving points 1', 2', and 3'. The distances " C "-1',
etc., are used to construct the elevation ellipse, as explained
for the previous example. For the lay-out construction the
projector lengths 1-1', 2-2', and 3-3' are marked off from 0
and 6 on the lay-out perpendicular. Also from 0 and 6 is
measured the half-base width A-S, and a line, perpendicular
to the first, drawn from the point and marked A, B, C, and D.
The pattern is drafted in the usual manner by obtaining the
true lengths of the elevation lines on the lay-out and using
these to construct the pattern triangles in conjunction with
the transformer top edge spacings and base distances.

An alternative method of tackling a twisted surface problem
is shown in Fig. 40, Chapter Ten. This method can be applied,
if desired, to those cases where an end view or plan of the job

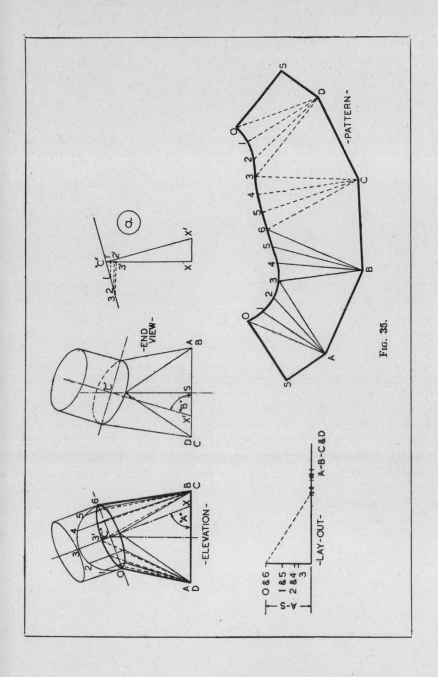

Fig. 35.

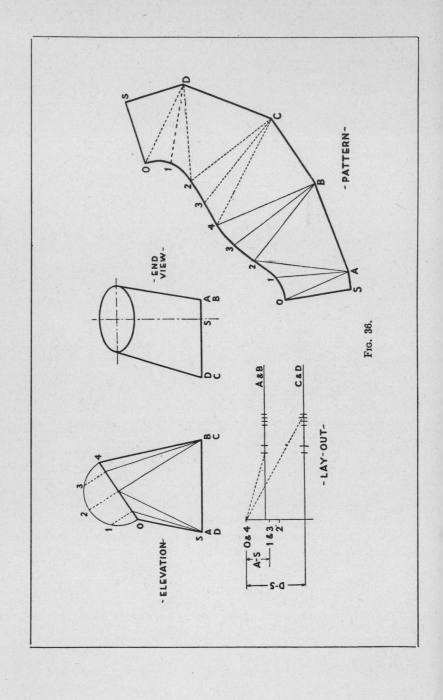

-PATTERN-

-END VIEW-

-ELEVATION-

-LAY-OUT-

Fig. 36.

is shown by the draughtsman in a different position to the examples already given. The craftsman who can read working readings and has a fair knowledge of geometry should find little difficulty in adapting to the solution of problems of this type the drafting methods described in this book, however the information required is set out on the blue-print.

OFF-CENTRE TRANSFORMER

In the problem illustrated in Fig. 36 the inclined round face of the transformer top is off-centre from its rectangular base. The vertical centre line in the end view can be assumed to be the edge of a cutting plane, and base distances each side of it are marked off in the lay-out, as explained for the example shown in Fig. 34.

The construction of the lay-out and drafting of the pattern should entail little trouble, the full lines in the pattern being those on the front of the transformer, and they are obtained from the lay-out horizontal line A and B, while the dotted lines are on the transformer back ; these true lengths are derived from C and D line in the lay-out.

Commence the pattern by drawing in a suitable position the elevation edge line 0-S. Next measure the same line along lay-out line A and B and join up to 0 and 4 on the perpendicular. With this true length describe an arc from point 0 on the pattern seam and cut it in A from centre S with distance A-S from the end view or lay-out. Take elevation length 1-A, measure it along lay-out line A and B, and join to 1 on the perpendicular.

Describe an arc with this true length from point A in the pattern and cut in 1 with spacing 0-1 from the elevation semi-circle. Take false length A-2 from the elevation, measure along lay-out line A, and connect to point 2 on the perpendicu-lar. Strike an arc with the true length from centre A in the pattern. Cut this arc in 2 with spacing 1-2. Next measure false length 2-B along lay-out line A and B and connect to 2 on the perpendicular. Strike an arc from point 2 in the pattern with the true length line and cut in B with elevation base distance A-B. Complete the remainder of the pattern on similar lines.

Chapter Ten

MISCELLANEOUS PROBLEMS

THE examples now to be described have several points of interest which show the adaptability of the lay-out system to the solution of various kinds of development problems.

Fig. 37 depicts the setting-out required for an air-duct connecting piece or shoe, fitting off-centre on to a round main. It will be seen from a study of the elevation that the job essentially consists of the intersection of a transformer of irregular shape with a cylinder. Before the patterns can be obtained the joint line between the parts must be found in the following manner :—

SETTING-OUT FOR OFF-CENTRE SHOE

First set-out an elevation and draw a half-section of the transformer base on the cylinder centre line, as shown. Divide each of the semicircular ends into two equal parts and draw perpendiculars to the centre line. Next describe a semicircle on the transformer top, divide it into four equal parts and draw in ordinates to the edge line. Connect the top and base edge points with straight lines as shown. Draw an end view, describe a semicircle on the top edge, and divide it into four equal parts. Number the points 0, 1, 2, 3, and 4 and draw perpendiculars to the top edge. Produce these lines to cut the cylinder circumference at points Bb, Cc, and Dd.

Now project all the points on the end view joint line across horizontally to the elevation to cut the lines previously drawn on the transformer surface at points A, B . . . b, a. Connect these points to obtain the required joint line. Next find the shape of the hole in the main pipe. Erect a vertical line in a convenient position and mark off on it spacings Aa-Bb, Bb-Cc, Cc-Dd, and Dd-Ee from the end view. Draw horizontal lines of indefinite length through the points. Make a-A, b-B, etc., equal in length each side of A-E to the similar lettered lines in the elevation. Join up the points to obtain the perimeter

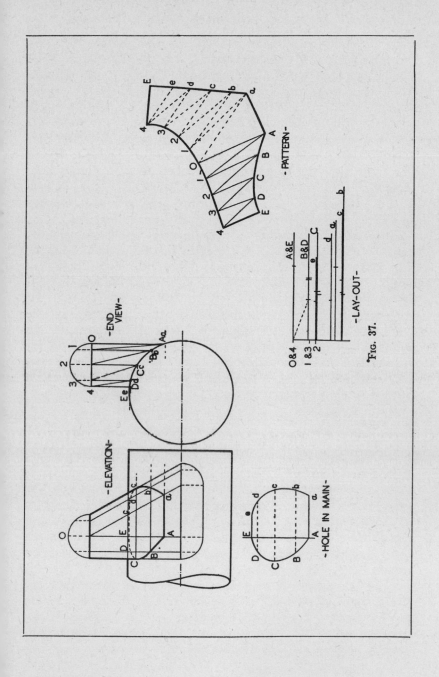

-END VIEW-

-PATTERN-

-LAY-OUT-

A & E
B & D
C
e
d
a
c
b

O & 4
1 & 3
2

-ELEVATION-

-HOLE IN MAIN-

Fig. 37.

of the hole, which is the true length of the joint line between the connecting piece and the main. The pattern for the shoe can be triangulated from the end view in the usual manner. To construct the lay-out draw a vertical line, mark a point 0 and 4, and step off distances 1 and 3 and 2 equal in length to the top-edge ordinates. Draw a horizontal line from 0 and 4 and letter it A and E. From 0 and 4 step off on the perpendicular the distances from the hole which lie each side of A-E, and draw horizontal lines from the points. Letter them as shown. Note that ordinates B and D are the same length in this example, thus the appropriate lay-out line is marked B and D.

No difficulty should be experienced in developing the pattern by obtaining the true lengths in the lay-out of all the lines drawn on the shoe surface in the end view. Each line actually represents two lines in the pattern. False length 2-Cc, for example, is first stepped off on lay-out line C, and triangulated to 2 on the perpendicular to obtain 2-C in the pattern. Later in the development it is used again by marking it off on lay-out line c and measuring the distance to 2 on the perpendicular to find pattern line 2-c. It is important to note that the spacings E-D, D-C, etc., in the pattern are taken from the hole perimeter and not from the elevation joint line.

SHOE FITTED TO CONICAL CONNECTING PIECE

In Fig. 38 the conical duct has a shoe fitted at right angles to its centre line, as shown. To find the joint line, strike arcs from the cone centre in the half-end view with distances b-b', c-c', and d-d' from the elevation (section line d-d' is drawn in any suitable position). These arcs are intersected at B, C, and D with the shoe edge line and a line drawn parallel to it from 1. Draw horizontal lines from the points to cut the elevation section lines at B, C, and D.

The lay-out is quite simple, as the horizontal lines are drawn from points 0 and 4, 1 and 3, and 2. Reference to the problem given in ·Fig. 31 should make the drafting of the quarter-pattern for the cone frustum clear, the shoe half-pattern being perfectly straightforward to develop.

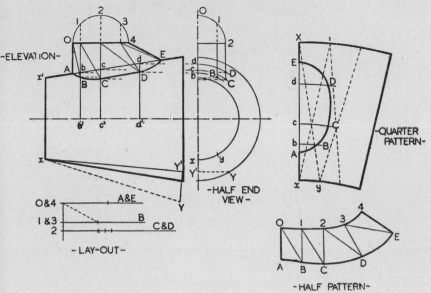

Fig. 38.

PETROL MEASURE TOP

This problem, as illustrated in Fig. 39, has a shaped top, the true perimeter of which is found by measuring off distances A-*b*, *b-c*, etc., along a straight line as at (*a*), erecting perpendiculars from the points and cutting these off the same length as ordinates B, C, D, E, and F on A-G. The spacings A-B, C-D, etc., at (*a*) are those used for the pattern. Part of this is slightly different to develop from previous problems. When finding the true length of D-K mark it along lay-out line K and triangulate to the point on the perpendicular made by the line D. Repeat this procedure for the remaining lines on this part of the job.

ELLIPTICAL HOOD TWISTED TO OUTLET PIPE

In this example, shown in Fig. 40, the end view vertical line *a*-A is divided into six suitable parts and horizontal lines drawn through the points across the end view surface to give points B, C, D, etc., on the perimeter. The lay-out is constructed as shown, the distances from 0 and 6 being the same as those which lie between vertical line *a*-A and the points

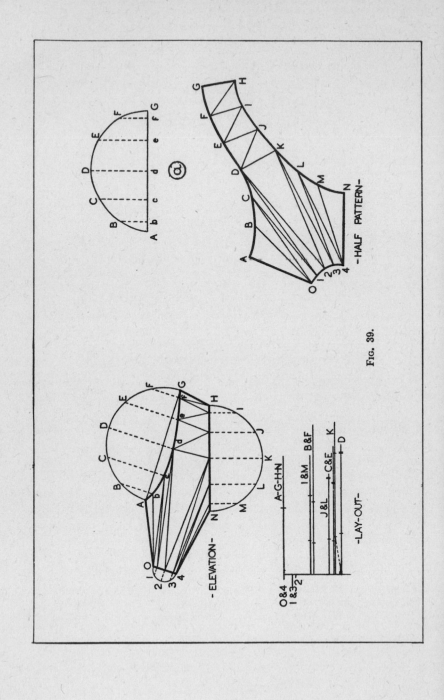

Fig. 39.

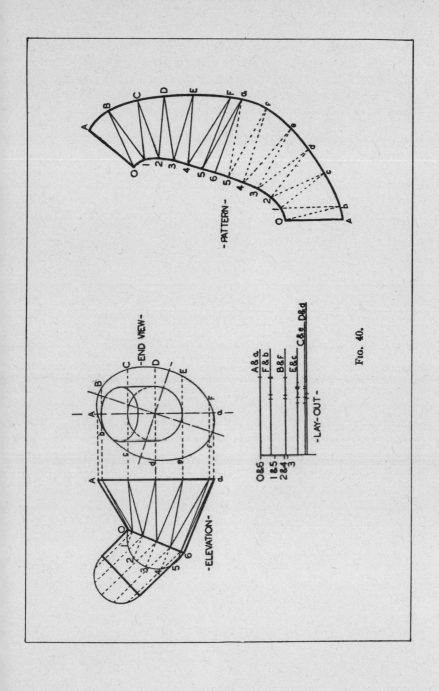

-PATTERN-

-END VIEW-

-ELEVATION-

O & 6	A & a
1 & 5	F & b
2 & 4	B & f
3	E & c
	C & e D & d

-LAY-OUT-

FIG. 40.

on the ellipse perimeter. It will be observed that ordinate
B is equal in length to ordinate f, C equal to e, and so on.

The construction of the lay-out and drafting of the pattern
is on similar lines to previous examples. The outlet pipe is
marked out by parallel-line development.

RECTANGULAR TRANSITION BEND

The example depicted in Fig. 41 is of a transition
bend between two rectangular ducts of different size. An

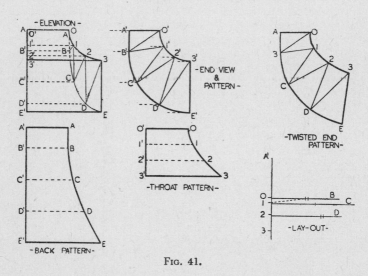

FIG. 41.

end view of the job is first drawn, and it is used as the pattern
for the flat side or cheek, in addition to being used as a basis
for triangulating the twisted-end pattern. The surface of the
end view is divided up into a suitable number of triangles,
and points 1′, 2′, 3′, B′, C′, and D′ projected horizontally
across to an elevation drawn as shown to obtain distances
D-D′, C-C′, etc. It is necessary to use the parallel-line method
of development to draft the back and throat patterns. For the
back pattern the stretch-out line A′ . . . E′ is made equal in
length to the curve A′ . . . E′ in the end view. Perpendiculars drawn from points A′, B′, etc., are made equal in
length to the distances A-A′, B-B′, etc., in the elevation.

The throat pattern is obtained similarly. For the twisted-end pattern it is necessary to construct a lay-out by drawing a vertical line, marking a point A' and measuring from it distances 0-0', 1-1', etc., from the elevation or throat pattern, marking the points 0, 1, 2, and 3. From A' is also measured off distances A-A', B-B', etc., from either the back pattern or elevation and horizontal lines drawn from the points and lettered B, C, and D respectively. The twisted-end pattern is triangulated from the end view by obtaining the true length lines in the lay-out and building up the pattern triangles with them in conjunction with the spacings from the curved lines in the throat and back patterns in the following manner.

First draw in a convenient position the top twisted-end pattern line 0-A equal in length to 0'-A' in the end view. Next take false length 0'-B' from the end view, measure it along lay-out line B and join to 0 on the perpendicular. Describe an arc from centre 0 on the pattern line with this true length. Cut this arc in B with spacing A-B from the back pattern. Next take 1'-B' from the end view, measure it along lay-out line B and triangulate to 1 on the perpendicular. From point B in the pattern strike an arc with the true length obtained and cut in 1 with spacing 0-1 from the throat pattern.

Now take distance 1'-C' from the end view, step it along lay-out line C, and join to 1 on the perpendicular. With this true length describe an arc from point 1 in the pattern, and cut it in C with the back pattern spacing B-C. Pick up 2'-C' from the end view, measure off on the lay-out line C, and connect to point 2 on the perpendicular. From point C in the pattern strike an arc with the true length and cut in 2 with throat spacing 1-2. Complete the pattern, using the same procedure for the remaining lines.

There are innumerable other articles which lend themselves to development by means of triangulation, but sufficient examples have been given to enable all those who have understood the principles of the lay-out system to practise with confidence the art of geometrical pattern drafting.

Chapter Eleven

ALLOWANCES FOR METAL THICKNESS

SHEET METAL patterns developed by geometrical methods are rarely very accurate, and a certain amount of arithmetic must be used to enable suitable adjustments to be made to the templates before they can be put into service.

Unless the metal used for the job is of very light gauge, allowances for material thickness must be taken into consideration when laying-out patterns. The principle on which metal thickness allowances are calculated is that of the " mean line " whereby the middle line of the metal is assumed to have not changed its length when the job is shaped and bent; this length, when obtained, is the required " stretched-out " length of the work in the flat sheet.

In Fig. 42 at (a) is shown a cylinder, the end view of which

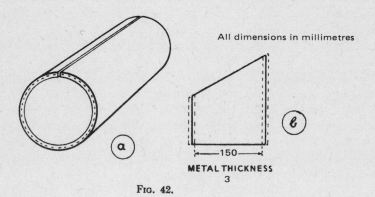

All dimensions in millimetres

METAL THICKNESS
3

FIG. 42.

depicts the mean or centre line of the metal thickness. The length of this dotted line can be calculated by obtaining the diameter of the middle circle and multiplying it by 3·1416. If the inside diameter of the cylinder is given, then the mean diameter is the inside diameter plus one thickness of metal. Thus if the given diameter is 150 mm and the metal thickness 3 mm, the length of the centre line for the pattern girth line would be 153 × 3·1416, which is equal to 480·66 mm. If the out-side diameter of the cylinder were given, the required length would be (the outside diameter minus the metal thickness multiplied by 3·1416).

CORRECT SETTING-OUT PATTERNS

When setting-out patterns for any job, the elevation or other views must always include the correct position of the mean line. The parallel edge lines of the cut round pipe, for instance, as shown in Fig. 2, p. 10, should actually represent the mean lines, as illustrated in Fig. 42 at (*b*). In this diagram dotted lines denote the metal thickness. As the dimensions of this pipe are the same as the cylinder given at (*a*), the stretch-out line for the pattern should be measured off 480 mm in length, which is correct enough for all practical purposes.

Similarly, the elevation of the cone frustum in Fig. 43 is drawn to the correct dimensions and the mean lines produced

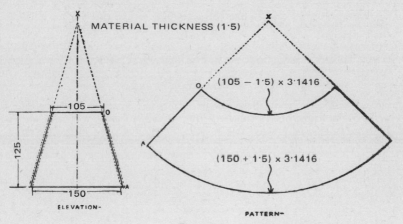

FIG. 43.   All dimensions in millimetres

to an apex to obtain the radius point for the pattern. The correct girth lines of the pattern are then calculated and measured off as shown. In actual practice it is the best plan when laying out a drawing to draw all edge lines in the positions which would be occupied by the mean lines in a fully dimensioned drawing as executed by a draughtsman.

Referring to Fig. 43, it would be necessary to draw the cone base line 151·5 mm long on the metal and the top edge line 103·5 mm long, thus obtaining the right positions for the slanting edge lines. This ensures the correct length for the pattern generator line when the edge lines are produced to an apex. If all drawings are set-out on this principle and the pattern girth lines checked by calculation of circumferences, etc., and increased in length if necessary, no difficulty should be experienced in obtaining really accurate patterns.

SQUARE AND RECTANGULAR WORK

To obtain the patterns for square or rectangular work, such as are used for ducts, it is often necessary to make allowances for the pipe ends to slip into each other to enable a riveted or

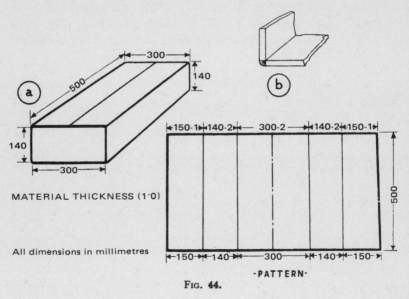

FIG. 44.

spot-welded seam to be made. The sides of the job, as shown in Fig. 44 (*a*), are folded up at right angles with fairly sharp inside corners. This means that no allowances for material thickness have to be made when working to inside dimensions because the metal stretches when it is folded, as shown by the shaded lines at Fig. 44 (*b*).

For the inside dimensions of the large end of the duct the measurements are marked off on the pattern base line, each side of the vertical centre line. The small or slip-in end of the pattern must have twice the thickness of metal deducted from each of the sides to give the required outside dimensions when the job is folded up. A careful study of Fig. 44 should make the whole procedure clear.

BEND ALLOWANCES

The patterns for aircraft detail fittings and similar jobs have to be very accurately developed, therefore " rule-of-thumb " methods are completely out of place in the modern factory. By using the principle of the mean line it is possible to mark out templates for work which must be made to fine limits. No sharp bends are allowed in aircraft work, and unless otherwise shown by the draughtsman it is the general rule to make all jobs such as brackets, clips, etc., with a bend inside radius of twice the thickness of metal.

To obtain the developed length of the fitting on the flat sheet it is necessary to find the length of the mean line by calculating the lengths of the flat portions and the bends separately. The stretched-out length of the bend, called the bend allowance, can be found by the following formula :—

Bend allowance = mean radius of bend multiplied by
bend angle multiplied by ·01745.

This formula can be applied to any type of bend, as it has been tried and proved on innumerable occasions in actual practice.

In Fig. 45 (a) is illustrated a right-angle bracket made of metal 1·5 mm thick. To obtain its development, first find the length of flat x. Deduct 3G—three times the metal thickness—from 55 mm, which gives the required dimension 50·5 mm. The bending allowance is the mean radius $(2G + \frac{1}{2}G)$ multiplied by 90 (the number of degrees in the bend angle) multiplied by ·01745. The B.A., or length of the bend mean line, is thus 5·89 mm. Flat y is 65 mm − 3G which is equal to 60·5 mm.

DEVELOPMENT OF PIPE CLIP

The clip shown at Fig. 45 (b) is made of metal 1 mm thick.

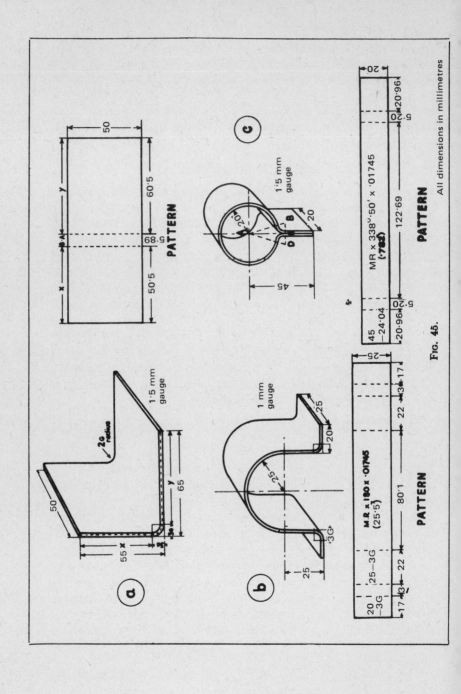

All dimensions in millimetres

Fig. 45.

a — 1·5 mm gauge, 2G radius, 50, 55 x, y, 65, 30 R

b — 1 mm gauge, 25, 25 R, 20, 25, 3G

c — 1·5 mm gauge, 45, 20, 20°, B, D

PATTERN (upper right): 50, x, 68·5, B A, 50·5, 60·5, y

PATTERN (b): M.R x 180 x ·01745 (25·5), 20, 3G, 17, 3, 22, 25-3G, 80·1, 22, 3, 17, 25

PATTERN (c): M R x 338·50′ x ·01745 (·782), 20, 5·0, 96·20 N, 122·69, 20·96, 5·0, 24·04, 45, 20·96, 5·0, 96·20 N

Essentially it is comprised of two similar right-angle brackets and a semicircular bend. Calculate the length of the bracket mean line by the method explained for the previous example. As there are 180° in the semicircular bend, the B.A. for this is found as follows :—

$$B.A. = M.R. \times 180 \times \cdot01745$$
$$= (25 + \cdot5G) \times 180 \times \cdot01745$$
$$= 25\cdot5 \times 180 \times \cdot01745$$
$$= 80\cdot10 \text{ mm.}$$

No further explanation should be necessary, if the diagram is carefully studied, to draw the full development on the flat sheet.

DEVELOPMENT OF TUBE FITTING

The fitting illustrated in Fig. 45 (c) is rather more difficult to develop, because the bend angles cannot be easily recognised, as in the case of the previous examples. To find the correct angles it is necessary to use trigonometry. Referring to the diagram it will be seen that the small bends are separated from the large bends by lines which join the radius points of the bends. These lines form the isosceles triangle BAD.

A perpendicular raised from C to A forms two right-angled triangles ABC and ADC. The length of hypotenuse AB and the base BC of triangle ABC can be obtained from the working drawing. BC divided by AC is the cosine of angle ABC.

To obtain the length AB add the inside radius of the large bend to the outside radius of the small bend (i.e. AB = 20 + (G + 2G) = 24·5 mm). BC is 3G in length. Thus the cosine of the required angle is found as follows :—

$$\cos ABC = \frac{4\cdot5}{24\cdot5} = \cdot1837$$

$$\therefore \text{ angle ABC} = 79° \ 25'.$$

The B.A. for each of the small bends is :—

$$B.A. = 2\cdot5G \times 79° \ 25' \times \cdot01745$$
$$= 3\cdot75 \times 79\cdot42 \times \cdot01745$$
$$= 5\cdot20 \text{ mm.}$$

To find the number of degrees in the large bend, first obtain angle BAC by deducting 79° 25′ from 90°. This gives 10° 35′, which must be doubled to obtain the top angle of the isosceles triangle. This 21° 10′ is deducted from 360° to give 338° 50′ as the large bending angle. The B.A. for this bend is :—

$$B.A. = M.R. \times 338° \; 50' \times \cdot01745$$
$$= (20 + \cdot75) \times 338 \cdot 83 \times \cdot01745$$
$$= 122 \cdot 69 \text{ mm.}$$

Next, obtain the length of the flat portions. First find the length of AC as follows :—

$$\tan 79° \; 25' = \frac{AC}{4 \cdot 5} \qquad \text{so } AC = \tan 79° \; 25' \times 4 \cdot 5$$

$$\therefore \; AC = 24 \cdot 04 \text{ mm.}$$

Finally, deduct 24·04 mm from the given dimension 45 mm to obtain 20·96 mm as the length of each flat.

All the foregoing principles can be applied to the solution of any development problem in which it is necessary to make accurate allowances for the thickness of sheet metal.

BEND ALLOWANCE TABLE

As the majority of bends are made at an angle of 90°, the following table of bend allowances should be of service. They are calculated for an inside bend radius of twice the metal thickness.

Material	Nearest Metric Thickness	Bend Allowance
10 S.W.G.	3 mm	11·78 mm
12 S.W.G.	2·5 mm	9·82 mm
14 S.W.G.	2 mm	7·85 mm
16 S.W.G.	1·6 mm	6·28 mm
18 S.W.G.	1·2 mm	4·71 mm
20 S.W.G.	0·9 mm	3·53 mm
22 S.W.G.	0·7 mm	2·75 mm
24 S.W.G.	0·6 mm	2·35 mm

Chapter Twelve

CALCULATION METHODS

ALTHOUGH geometrical methods of construction for the solution of problems of pattern development are universally used in the sheet metal industry, the results obtained are not always accurate enough for precision work. As an alternative, mathematical formulæ can be used to obtain patterns within reasonable dimensional limits in those cases where the surface of the job lends itself to be calculated.

When the size of the article is large a distinct advantage is that, as the reproduction of full-size plans, elevations, or other views from the working drawing on to the sheet metal is not required, much valuable time can be saved, together with economy of material.

The advantage of the mathematical approach is that the true lengths of pattern lines can be calculated from the given dimensions of the job by the use of appropriate formulæ. Several such formulæ will be presented and proved in subsequent chapters of this book, and it is hoped that their obvious convenience will encourage more widespread use of calculation methods. The problems themselves can be readily solved with the aid of tables, of the slide rule, or even of a pocket electronic calculator if one is available having the necessary range.

In each of the examples explained, the *most accurate formula possible* is given, taking into account the thickness of the material being used for the job in question. Accuracy of this order is not always necessary, however; and in a great many operations the formulæ given at the foot of page 100, in which the *overall radius* is used in place of the *mean radius,* will be found perfectly adequate, and will certainly make much of the arithmetic involved a good deal easier to work out.

In each case, however, it is advisable that the formulæ for obtaining the girths of the pattern be used, as it is necessary to take cognisance of the metal thickness to obtain the correct finished dimensions of the end surfaces of the article.

METHOD OF TRIANGULATION

As previously explained, the triangulation method of pattern development is extensively used in sheet metal work. The essential principle of this is dividing the plan or elevation of an object into a suitable number of triangles, finding the true lengths of the sides, and laying-out with them pattern triangles in a consecutive manner on to the flat sheet. There are several geometrical methods of construction to obtain the true lengths of lines for the pattern triangles, using the plan, or, as in the lay-out system, the elevation of the object as a basis.

The correct length of a development line on the surface of an object can be calculated when its projected length on a contiguous vertical plane is obtainable. If the projected or elevation length is made the base of a right-angled triangle, of which the perpendicular is constructed from the difference in length of the projectors, the hypotenuse will be the true length of the elevation line.

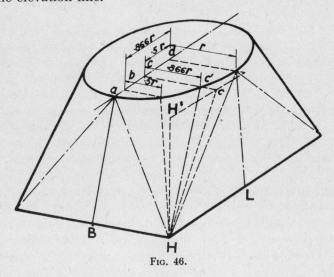

FIG. 46.

In Fig. 46 is given a perspective view of a rectangular to round transformer, which illustrates the principle of finding the true lengths of lines from an orthographic projection as shown on a working drawing. A plan and elevation of the job are shown in Fig. 47. Chain-dotted lines on the elevation represent the projections of lines on the surface of the job. Fig. 47 shows a half-pattern of the transformer in which true length lines Ba, Ha, Hb, Hc, Hd, and Ld are used to construct pattern triangles.

Referring again to Fig. 46, it can be seen that development lines are drawn to corner H from equally spaced points around the top edge. From these points dotted lines are shown projected perpendicularly back to the top centre line. The length of ordinate b is ·5 of the radius, and ordinate c ·866 of the radius, while ordinate d is the radius. Ordinates c and d are ·5 of the radius apart, and ordinates b and d have a distance between them of ·866 of the radius. From these dimensions the lengths of projectors from the edge points to a vertical plane erected on base line HL can be calculated. Having ascertained the position and length of the projector lines, the pattern true lengths can be calculated by an adaptation of the theorem of Pythagoras.

In Fig. 46 the method of obtaining the true length of one of the pattern lines Hc' is shown. Hc is the hypotenuse of a right-angled triangle, of which the perpendicular is the height of the job HH'. The base length H'c is obtained by deducting ·5 of the radius from the base edge distance HL. Projector length cc' is found by deducting ordinate c, which is ·866 of the radius, from the base half-width BH. Extracting the root of the sum of the squares on HH', H'c and cc' gives the true length Hc'.

From the foregoing it should be apparent that true length Hc' can be found by using the formula :—

$$\mathrm{H}c' = \sqrt{(\mathrm{HL} - ·5 \text{ top rad.})^2 + (\mathrm{HB} - ·866 \text{ top rad.})^2 + \text{height}^2}.$$

It is only necessary to calculate true length lines for a quarter of the pattern, as the top and bottom of the transformer are symmetrical about common centre lines.

RECTANGULAR TO ROUND TRANSFORMER

The example given is of a transformer made of 1 mm thick material, the top of the transformer being 200 mm inside diameter, and the base 300×250 mm, also inside sizes. In each of the formulæ mr denotes the mean radius. For the given example the mean radius, which is the radius plus half the metal thickness, is $100 + 0·5 = 100·5$ mm.

Note that as the inside sizes for the base are given, these lengths are used for the pattern because, as the transformer has more or less sharp inside corners, these dimensions will be sufficiently accurate.

The lengths of chords ab, bc, etc., in the pattern are found by multiplying the mean radius of the top by ·522. Therefore ab is $100·5 \times ·522 = 52·46$ mm. The correct constant for a twelfth part of the top circumference is ·5176. It is advisable, however, to use ·522 times the mean radius for the length of pattern chords, as the arcs which are drawn in freehand between points a and b, b and c, etc., are of larger radius than that of the transformer top, and the use of a slightly longer length of chord gives a pattern arc which is practically the correct size. A check can be made by measuring the girth line of the completed half-pattern with the distance found by multiplying the mean radius by 3·1416.

FORMULÆ FOR OBTAINING TRUE LENGTH LINES

By using the following formulæ each of the pattern true length lines can be calculated from the given dimensions in Fig. 47 :—

True length Ba

$$= \sqrt{(\text{HL} - \text{top radius})^2 + \text{height}^2}$$
$$= \sqrt{(150 - 100)^2 + 190^2}$$
$$= 196·47 \text{ mm.}$$

True length Hb

$$= \sqrt{(\text{HL} - ·866r)^2 + (\text{HB} - ·5r)^2 + h^2}$$
$$= \sqrt{(150 - ·866 \times 100)^2 \times (125 - ·5 \times 100)^2 + 190^2}$$
$$= 213·88 \text{ mm.}$$

True length Hc

$$= \sqrt{(\text{HL} - ·5r)^2 + (\text{HB} - ·866r)^2 + h^2}$$
$$= \sqrt{(150 - ·5 \times 100)^2 + (125 - ·866 \times 100)^2 + 190^2}$$
$$= 218·12 \text{ mm.}$$

True length Hd

$$\sqrt{\text{HL}^2 + (\text{H}b - r)^2 + h^2}$$
$$= \sqrt{150^2 + (125 - 100)^2 + 190^2}$$
$$= 243·36 \text{ mm.}$$

True length Ld

$$= \sqrt{(HB - r)^2 + h^2}$$
$$= \sqrt{(125 - 100)^2 + 190^2}$$
$$= 191 \cdot 64 \text{ mm.}$$

The true lengths obtained can be used to lay-out the pattern, but if more accurate results are desired the mean radius of the top must be taken into account in the formulæ in the following manner :—

True length Ba

$$= \sqrt{(HL - \text{top mean radius})^2 + \text{height}^2}$$
$$= \sqrt{(150 - 100 \cdot 5)^2 + 190^2}$$
$$= 196 \cdot 34 \text{ mm.}$$

True length Hb

$$= \sqrt{(HL - \cdot 866mr)^2 + (HB - \cdot 5mr)^2 + h^2}$$
$$= \sqrt{(150 - \cdot 866 \times 100 \cdot 5)^2 + (125 - \cdot 5 \times 100 \cdot 5) + 190^2}$$
$$= 213 \cdot 74 \text{ mm.}$$

True length Hc

$$= \sqrt{(HL - \cdot 5mr)^2 + (HB - \cdot 866mr)^2 + h^2}$$
$$= \sqrt{(150 - \cdot 5 \times 100 \cdot 5)^2 + (125 - \cdot 866mr)^2 + 190^2}$$
$$= 217 \cdot 92 \text{ mm.}$$

True length Hd

$$= \sqrt{HL^2 + (HB - mr)^2 + h^2}$$
$$= \sqrt{150^2 + (125 - 100 \cdot 5)^2 + 190^2}$$
$$= 243 \cdot 31 \text{ mm.}$$

True length Ld

$$= \sqrt{(HB - mr)^2 + h^2}$$
$$= \sqrt{(125 - 100 \cdot 5)^2 + 190^2}$$
$$= 191 \cdot 57 \text{ mm.}$$

It will be found a great aid to quick working to begin by calculating the values of the mean radius of the job to be tackled, and of ·866 and ·5 times this mean radius. These expressions recur in most of the formulæ, and once their values are known it is an easy matter to substitute them in the formula

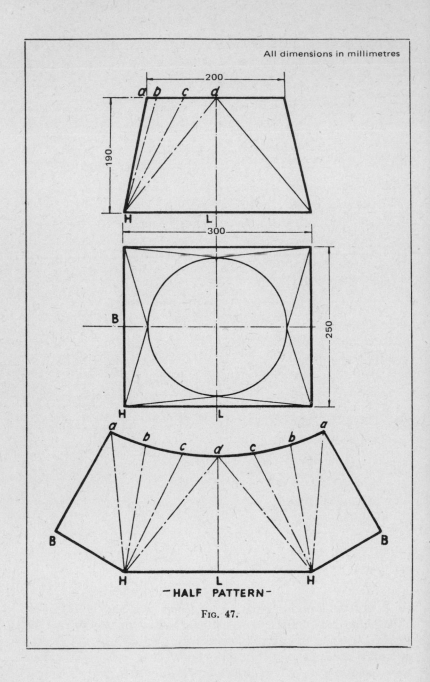

All dimensions in millimetres

−HALF PATTERN−

Fig. 47.

before working the problem out. This method of working will greatly facilitate the task of calculation in most of the examples given in later chapters of this book.

It must be clearly understood that all the chain-dotted construction lines shown on the elevation of this and subsequent examples are for explanatory purposes only. Guidance for the laying-out of the jobs on to the sheet metal must be obtained from the pattern diagrams.

DESCRIPTION OF PATTERN LAY-OUT

Draft the half-pattern as in Fig. 47 as follows :—

Draw line Ba (196·34 mm) in a convenient position at one edge of the sheet metal, and from B draw a perpendicular of indefinite length. Cut this off 125 mm in length, obtaining point H. Join H to a, so making unnecessary the calculation of the length of this line. With H as centre describe an arc with true length Hb (213·74 mm), and cut this arc in b with distance ab (52·46 mm).

Next from centre H describe true length Hc (217·92 mm) and cut with distance bc. Also from centre H describe true length Hd (243·31 mm), and cut the arc with cd from c. Take distance Ld and, from d as centre, cut the arc in L with distance HL (150 mm). This completes the quarter-pattern.

Next, with centre d and distance Hd, strike an arc, and from centre L, cut in H with distance HL (150 mm). Take distance Hc and from centre H draw an arc. Cut this arc in c with distance dc. Also from centre H mark off distance Hb and cut in b with cb. Next strike distance Ha from H, and cut in a with ba. From centre a describe an arc with distance Ba. Finally, cut this arc in B with base distance HB from centre H. Join all the top points with a smooth curve to complete the half-pattern.

THE POTENTIALITIES OF CALCULATION METHODS

The foregoing method of obtaining formulæ for pattern developments by triangulation can be applied to many and varied objects. It is not practical in the scope of this work to give suitable formulæ for every type of article, but sufficient examples will be worked out to give at least some idea of the potentialities of the subject. For instance, for repetition work it would be possible for working drawings to be issued with developments of the jobs given in addition to the usual views, as by using suitable formulæ the draughtsman could indicate the true lengths of all pattern lines.

Reproduction of these lines to their correct lengths on the sheet metal by the pattern drafter would enormously simplify his job and solve many of the problems so often encountered when laying-out developments in the shops.

Chapter Thirteen

OFF-CENTRE TRANSFORMER CALCULATIONS

PATTERNS for off-centre square or rectangular to round transformers can be calculated by using formulæ which, although similar to those given for the previous example, take into consideration the off-centre distances shown in plan.

OFF-CENTRE RECTANGULAR TO ROUND TRANSFORMERS

The next example, shown in Fig. 48, is of an off-centre transformer, 120 mm top inside diameter; base, 200 × 150 mm inside; depth, 130 mm; made from 1 mm thick material. As the job is symmetrical about the plan horizontal centre line, a half-pattern only need be drafted. Calculate the true lengths as follows :—

True length Ba

$$= \sqrt{(HL - mr)^2 + h^2} \qquad (HL = 200 - 125)$$
$$= \sqrt{(75 - 60 \cdot 5)^2 + 130^2}$$
$$= 130 \cdot 79 \text{ mm}$$

True length Hb

$$= \sqrt{(HL - \cdot 866mr)^2 + (HB - \cdot 5mr)^2 + h^2}$$
$$= \sqrt{(75 - 52 \cdot 39)^2 + (75 - 30 \cdot 25)^2 + 130^2}$$
$$= 139 \cdot 32 \text{ mm.}$$

True length Hc

$$= \sqrt{(HL - \cdot 5mr)^2 + (HB - \cdot 866mr)^2 + h^2}$$
$$= \sqrt{(75 - 30 \cdot 25)^2 + (75 - 52 \cdot 39)^2 + 130^2}$$
$$= 139 \cdot 32 \text{ mm.}$$

True length Hd

$$= \sqrt{HL^2 + (HB - mr)^2 + h^2}$$
$$= \sqrt{75^2 + (75 - 60 \cdot 5)^2 + 130^2}$$
$$= 150 \cdot 77 \text{ mm.}$$

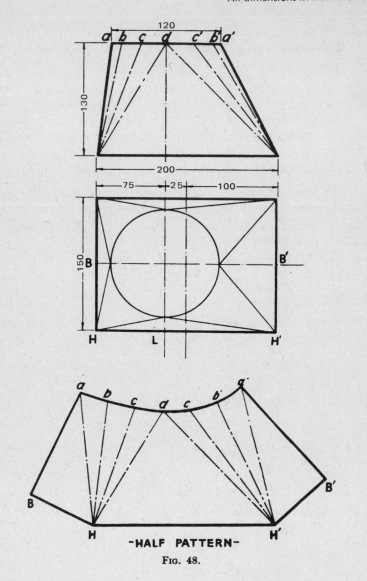

-HALF PATTERN-

Fig. 48.

True length H'd

$$= \sqrt{H'L^2 + (H'B' - mr)^2 + h^2} \qquad (H'L = 200 - 75)$$
$$= \sqrt{125^2 + (75 - 60 \cdot 5)^2 + 130^2}$$
$$= 180 \cdot 91 \text{ mm.}$$

True length H'c'

$$= \sqrt{(H'L - \cdot 5mr)^2 + (H'B' - \cdot 866mr)^2 + h^2}$$
$$= \sqrt{(125 - 30 \cdot 25)^2 + (75 - 52 \cdot 39)^2 + 130^2}$$
$$= 162 \cdot 42 \text{ mm.}$$

True length H'b'

$$= \sqrt{(H'L - \cdot 866mr)^2 + (H'B' - \cdot 5mr)^2 + h^2}$$
$$= \sqrt{(125 - 52 \cdot 39)^2 + (75 - 30 \cdot 25)^2 + 130^2}$$
$$= 155 \cdot 49 \text{ mm.}$$

True length H'a'

$$= \sqrt{(H'L - mr)^2 + H'B'^2 + h^2}$$
$$= \sqrt{(125 - 60 \cdot 5)^2 + 75^2 + 130^2}$$
$$= 163 \cdot 35 \text{ mm.}$$

True length B'a'

$$= \sqrt{(H'L' - mr)^2 + h^2}$$
$$= \sqrt{(125 - 60 \cdot 5)^2 + 130^2}$$
$$= 145 \cdot 11 \text{ mm.}$$

Length of chords ab, bc, etc. $= mr \times \cdot 522$
$$= 31 \cdot 58 \text{ mm.}$$

Draft the first quarter of the pattern, as in Fig. 48, in the same manner as described for the previous example. Note that there is no need to calculate the length of Ha, as HB is drawn at right angles to Ba. After marking off line Hd describe an arc from centre d with length H'd (180·91 mm). Cut this arc in H' from centre H with distance HH' (200 mm). From centre H' describe an arc with true length H'c (162·42 mm), and cut this arc in c' from centre d with chord dc' (31·58 mm). Continue the remainder of the development as shown.

Calculation methods as explained for the two previous examples can also be applied to the development of a pattern for the type of transformer as illustrated in Fig. 49. The top of this transformer is 550 mm inside diameter, with a base 900 × 800 mm inside, and a vertical height of 750 mm. It is made from 1 mm thick material.

In the plan the centre lines of the top circle are shown 75 mm offset each way from the base centre lines. The formulæ for obtaining the pattern lines are exactly as for Fig. 46, the construction being the same for each quarter of the job, but involving different dimensions. Note that it is the centre lines of the top which determine points L, l, B, and B′, and base distances on the plan profile.

Calculate the pattern lengths as follows:—

True length Ba

$$= \sqrt{(\mathrm{HL} - mr)^2 + h^2}$$
$$= \sqrt{(375 - 225 \cdot 5)^2 + 750^2}$$
$$= 764 \cdot 7 \text{ mm.}$$

True length Hb

$$= \sqrt{(\mathrm{HL} - \cdot 866mr)^2 + (\mathrm{HB} - \cdot 5mr)^2 + h^2}$$
$$= \sqrt{(375 - 195 \cdot 28)^2 + (475 - 112 \cdot 75)^2 + 750^2}$$
$$= 852 \cdot 1 \text{ mm.}$$

True length Hc

$$= \sqrt{(\mathrm{HL} - \cdot 5mr)^2 + (\mathrm{HB} - \cdot 866mr)^2 + h^2}$$
$$= \sqrt{(375 - 112 \cdot 75)^2 + (475 - 195 \cdot 28)^2 + 750^2}$$
$$= 842 \cdot 3 \text{ mm.}$$

True length Hd

$$= \sqrt{\mathrm{HL}^2 + (\mathrm{HB} - mr)^2 + h^2}$$
$$= \sqrt{375^2 + (475 - 225 \cdot 5)^2 + 750^2}$$
$$= 874 \cdot 8 \text{ mm.}$$

True length H′d

$$= \sqrt{\mathrm{H'L}^2 + (\mathrm{H'B'} - mr)^2 + h^2}$$
$$= \sqrt{525^2 + (475 - 225 \cdot 5)^2 + 750^2}$$
$$= 948 \cdot 9 \text{ mm.}$$

True length H′c

$$= \sqrt{(\mathrm{H'L} - \cdot 5mr)^2 + (\mathrm{H'B'} - \cdot 866mr)^2 + h^2}$$
$$= \sqrt{(525 - 112 \cdot 75)^2 + (475 - 195 \cdot 28)^2 + 750^2}$$
$$= 900 \cdot 4 \text{ mm.}$$

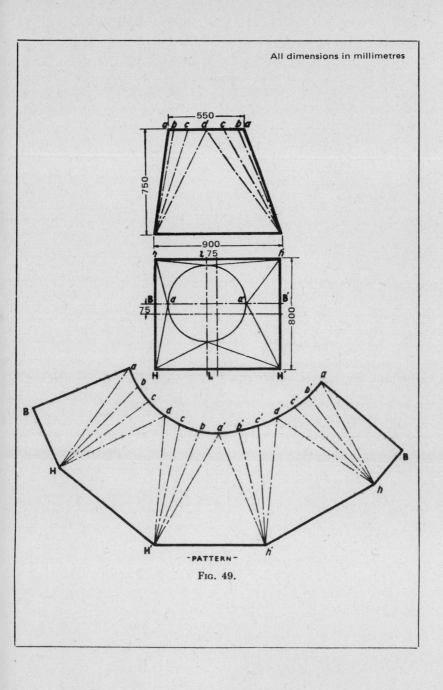

All dimensions in millimetres

-PATTERN-

FIG. 49.

True length H′b

$$= \sqrt{(\mathrm{H'L} - \cdot 886mr)^2 + (\mathrm{H'B'} - \cdot 5mr)^2 + h^2}$$
$$= \sqrt{(525 - 195 \cdot 28)^2 + (475 - 112 \cdot 75)^2 + 750^2}$$
$$= 895 \cdot 8 \text{ mm.}$$

True length H′a

$$= \sqrt{(\mathrm{H'L} - mr)^2 + \mathrm{H'B'}^2 + h^2}$$
$$= \sqrt{(525 - 225 \cdot 5)^2 + 475^2 + 750^2}$$
$$= 936 \cdot 9 \text{ mm.}$$

True length $h'a'$

$$= \sqrt{(h'l - mr)^2 + h'\mathrm{B'}^2 + h^2}$$
$$= \sqrt{(525 - 225 \cdot 5)^2 + 325^2 + 750^2}$$
$$= 870 \cdot 5 \text{ mm.}$$

True length $h'b'$

$$= \sqrt{(h'l - \cdot 866mr)^2 + (h'\mathrm{B'} - \cdot 5mr)^2 + h^2}$$
$$= \sqrt{(525 - \cdot 195 \cdot 28)^2 + (325 - 112 \cdot 75)^2 + 750^2}$$
$$= 846 \cdot 3 \text{ mm.}$$

True length $h'c'$

$$= \sqrt{(h'l - \cdot 5mr)^2 + (h'\mathrm{B'} - \cdot 866mr)^2 + h^2}$$
$$= \sqrt{(525 - 112 \cdot 75)^2 + (325 - 195 \cdot 28)^2 + 750^2}$$
$$= 865 \cdot 6 \text{ mm.}$$

True length $h'd'$

$$= \sqrt{(h'l)^2 + (h'\mathrm{B'} - mr)^2 + h^2}$$
$$= \sqrt{525^2 + (325 - 225 \cdot 5)^2 + 750^2}$$
$$= 920 \cdot 9 \text{ mm.}$$

True length hd'

$$= \sqrt{hl^2 + (h\mathrm{B} - mr)^2 + h^2}$$
$$= \sqrt{375^2 + (325 - 225 \cdot 5)^2 + 750^2}$$
$$= 844 \cdot 1 \text{ mm.}$$

True length hc'

$$= \sqrt{(hl - \cdot 5mr)^2 + (h\mathrm{B} - \cdot 866mr)^2 + h^2}$$
$$= \sqrt{(375 - 112 \cdot 75)^2 + (325 - 195 \cdot 28)^2 + 750^2}$$
$$= 805 \cdot 0 \text{ mm.}$$

True length hb'

$$= \sqrt{(hl - \cdot866mr)^2 + (h\mathrm{B} - \cdot5mr)^2 + h^2}$$
$$= \sqrt{(375 - 195\cdot28)^2 + (325 - 112\cdot75)^2 + 750^2}$$
$$= 799\cdot9 \text{ mm.}$$

True length ha

$$= \sqrt{(hl - mr)^2 + h\mathrm{B}^2 + h^2}$$
$$= \sqrt{(375 - 225\cdot5)^2 + 325^2 + 750^2}$$
$$= 831\cdot1 \text{ mm.}$$

Length of chords ab, bc, etc. $= mr \times \cdot522$
$$= 225\cdot5 \times \cdot522$$
$$= 117\cdot7 \text{ mm.}$$

DEVELOPMENT OF THE PATTERN

Mark off true length Ba (764·7 mm) in a convenient position and draw a perpendicular from B. Measure off a distance HB (475mm) and from H draw a line to a to complete the first pattern triangle. Next, from centre H, strike an arc, with true length Hb (852·1 mm), and cut it in b with chord ab (117·7 mm). Also from centre H describe an arc with Hc (842·3 mm), and cut it in c from centre b with spacing bc. Add true length Hd (874·8 mm) to the pattern in a similar manner, obtaining point d.

From the latter point describe an arc with true length H'd (948·9 mm). Cut this arc in H' with distance HH' (900 mm). Using H' as centre, mark off lines H'c, H'b, and H'a', and measure off spacings dc, cb, and ba', as shown. Continue the development in a similar fashion until complete, and finally join all the top points with a smooth curve.

RECTANGULAR TO ROUND TRANSFORMER BETWEEN OBLIQUE PLANES

Although there are many different kinds of air-duct transformers, the patterns for most of them can be readily calculated if the base and top lie between parallel planes. In cases where the top and the base of the transformer are between oblique planes, as in the example given in Fig. 50, more calculations are required to develop the pattern, because two vertical heights of the job must be taken into consideration. If only one height is shown on the working drawing, it becomes necessary to calculate the second height.

The transformer, which is made of 1 mm thick material, has a round top 280 mm inside diameter and a rectangular base 460×400 mm (inside sizes), which is inclined at an angle of $60°$ to the elevation vertical centre line. The overall vertical height is shown as 450 mm, but as the length of h' is not shown it must be calculated from the given dimensions.

Find the length of h' as follows:—

$$h' = h - (\cos 60° \times 460)$$
$$= 450 - (\cdot 5 \times 460)$$
$$= 220 \text{ mm.}$$

Next find base length HL:—

$$\text{HL} = \sin 60° \times \frac{460}{2}$$
$$= \cdot 866 \times 230$$
$$= 199 \cdot 18 \text{ mm.}$$

Calculate the true lengths as follows:—

True length Ba
$$= \sqrt{(\text{HL} - mr)^2 + h^2}$$
$$= \sqrt{(199 \cdot 18 - 140 \cdot 5)^2 + 450^2}$$
$$= 453 \cdot 8 \text{ mm.}$$

True length Hb
$$= \sqrt{(\text{HL} - \cdot 866mr)^2 + (\text{HB} - \cdot 5mr)^2 + 450^2}$$
$$= \sqrt{(199 \cdot 18 - 121 \cdot 67)^2 + (200 - 70 \cdot 25)^2 + 450^2}$$
$$= 474 \cdot 7 \text{ mm.}$$

True length Hc
$$= \sqrt{(\text{HL} - \cdot 5mr)^2 + (\text{HB} - \cdot 866mr)^2 + 450^2}$$
$$= \sqrt{(199 \cdot 18 - 70 \cdot 25)^2 + (200 - 121 \cdot 67)^2 + 450^2}$$
$$= 474 \cdot 6 \text{ mm.}$$

True length Hd
$$= \sqrt{\text{HL}^2 + (\text{HB} - mr)^2 + h^2}$$
$$= \sqrt{199 \cdot 18^2 + (200 - 140 \cdot 5)^2 + 450^2}$$
$$= 495 \cdot 7 \text{ mm.}$$

True length H'd

$$= \sqrt{H'L^2 + (H'B' - mr)^2 + h'^2}$$
$$= \sqrt{199 \cdot 18^2 + (200 - 140 \cdot 5)^2 + 220^2}$$
$$= 302 \cdot 7 \text{ mm.}$$

True length H'c'

$$= \sqrt{(H'l - \cdot 5mr)^2 + (H'B' - \cdot 866mr)^2 + h'^2}$$
$$= \sqrt{(199 \cdot 18 - 70 \cdot 25)^2 + (200 - 121 \cdot 67)^2 + 220^2}$$
$$= 266 \cdot 7 \text{ mm.}$$

True length H'b'

$$= \sqrt{(H'l - \cdot 866mr)^2 + (H'B' - \cdot 5mr)^2 + h'^2}$$
$$= \sqrt{(199 \cdot 18 - 121 \cdot 67)^2 + (200 - 70 \cdot 25)^2 + 220^2}$$
$$= 266 \cdot 9 \text{ mm.}$$

True length H'a'

$$= \sqrt{(H'l - mr)^2 + H'B'^2 + h'^2}$$
$$= \sqrt{(199 \cdot 18 - 140 \cdot 5)^2 + 200^2 + 220^2}$$
$$= 303 \cdot 1 \text{ mm.}$$

True length B'a'

$$= \sqrt{(H'L - mr)^2 + h'^2}$$
$$= \sqrt{(199 \cdot 18 - 140 \cdot 5)^2 + 220^2}$$
$$= 208 \cdot 4 \text{ mm.}$$

$$\text{Length of chords } ab, bc, \text{ etc.} = mr \times \cdot 522$$
$$= 140 \cdot 5 \times \cdot 522$$
$$= 73 \cdot 3 \text{ mm.}$$

No difficulty should be met with if the half-pattern development shown in Fig. 50 is followed.

LAYING-OUT THE PATTERN

Draw in a convenient position true length Ba (453·8 mm) and at right angles to it draw HB (200 mm) from B. Join H to a to complete the first triangle. Take true length Hb (474·7 mm) and describe an arc from centre H. Cut the arc in b from centre a with chord ab (73·3 mm). From H also describe an arc, this time with true length Hc (474·6 mm), and cut in c from centre b with bc. Next describe an arc from centre H

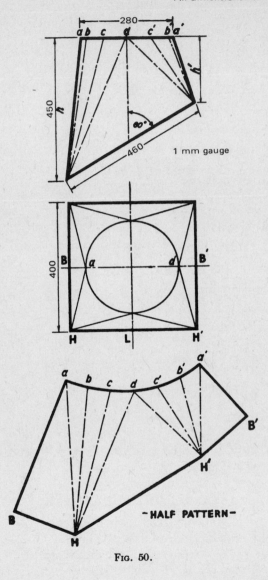

1 mm gauge

-HALF PATTERN-

Fig. 50.

with true length H*d* (495·7 mm) and cut in *d* from centre *c* with *cd*.

Next take true length H'*d* (302·7 mm) and from centre *d* describe an arc. With distance HH' (460 mm) from the elevation, cut this arc in H' from centre H. From centre H' describe an arc with true length H'*c*' (266·7 mm) and cut in *c*' from centre *d* with *c*'*d*. Now from centre H' strike an arc with true length H'*b*' (266·9 mm) and cut in *b*' from *c*' with *b*'*c*'. Take H'*a*' (303 mm) and describe an arc from H'. Cut this in *a*' from *b*' with *a*'*b*'.

From centre *a*' describe an arc with B'*a*' (208·4 mm) and cut in B' from centre H' with distance H'B' (200 mm) to complete the half-pattern.

Chapter Fourteen

CONICAL FRUSTUM CALCULATIONS

WHEN it is required to develop the pattern for a cone frustum of slight taper it is sometimes difficult to produce the elevation slanting edge lines to an apex as required for the radial-line system of development, so the triangulation method should be used. A cone frustum of this type is shown in Fig. 51. It has a top 650 mm outside diameter, a base 1000 mm outside diameter, and a vertical height of 750 mm. Material used is 1·5 mm thick.

It is only necessary to calculate two true length lines Aa and Ba for the development of the pattern. In the transformer examples all the true length lines lie on the surface of the metal when the jobs are shaped up, but in the type of article now described, line Ba actually is placed crosswise on the curved surface of the frustum, as it connects point B on the base edge of the cone to point a on the top edge.

For the most irregular-shaped articles it is a difficult process to obtain the exact length of curved connecting lines, and using geometrical methods it is always the practice to treat such lines as straight ones. In the case of right cone frustums of small taper, however, it is a simple matter to calculate the slightly increased length required for reasonable accuracy by increasing the distance between ordinates of ·866 mean radius to ·872 mean radius in the formula for obtaining the true length of Ba.

Thus the true length of Ba is calculated as follows, bearing in mind that MR is the mean radius of the base.

$$Ba = \sqrt{(\cdot872\text{MR} - mr)^2 + (\cdot5\text{MR})^2 + h^2}$$
$$= \sqrt{(\cdot872 \times 499\cdot25 - 324\cdot25)^2 + (\cdot5 \times 499\cdot25)^2 + 750^2}$$
$$= 798\cdot1 \text{ mm.}$$

$$Aa = \sqrt{(\text{MR} - mr)^2 + h^2}$$
$$= \sqrt{(499\cdot25 - 324\cdot25)^2 + 750^2}$$
$$= 770\cdot1 \text{ mm.}$$

All dimensions in millimetres

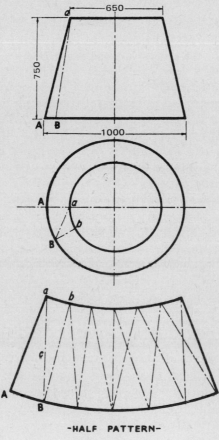

650

750

A B

1000

A a

b

B

a b

c

A

B

-HALF PATTERN-

Fig. 51.

$$\text{Chord AB} = \text{MR} \times \cdot522$$
$$= 499\cdot25 \times \cdot522$$
$$= 260\cdot6 \text{ mm.}$$
$$\text{Chord } ab = mr \times \cdot522$$
$$= 324\cdot25 \times \cdot522$$
$$= 169\cdot3 \text{ mm.}$$

PATTERN CONSTRUCTION

Draft the pattern by drawing Aa (770·1 mm) in a convenient position. From a describe an arc the true length Ba (798·1 mm). Cut this arc in B from centre A with chord AB (260·6 mm). Next, from centre B describe an arc with length Bb, which is the same length as Aa. Cut this arc in b with chord ab (169·3 mm) from centre a. Continue the development on the same lines for one quarter of the pattern.

The remaining quarter-pattern shows a quicker method of development. True length Ba is used to strike arcs alternately from points a, A, b, and B, thus eliminating the use of line Aa, except for the seams. This means that one setting of the trammels only is needed, after marking off Aa, when drafting the pattern. Although a half-pattern is shown, a full pattern can be made using the same methods.

CALCULATION OF HALF-PATTERN—SECOND METHOD

It is possible to calculate the cone frustum half-pattern by using other mathematical formulæ. Fig. 52 illustrates a method of obtaining the pattern development by calculating the pattern radius CC′, and the angle between the slanting edge lines of the half-pattern.

First of all calculate the length of pattern radius:—

$$\text{Pattern radius} = \frac{\text{MR} \times \text{AC}}{\text{MR} - mr}$$

$$= \frac{499\cdot25 \times 770\cdot1}{499\cdot25 - 324\cdot25}$$

$$= 2196\cdot9 \text{ mm.}$$

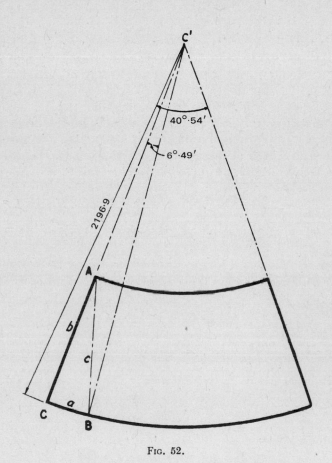

Fig. 52.

$$\text{Half pattern angle} = \frac{\text{MR} \times 180°}{\text{Pattern radius}}$$

$$= \frac{499 \cdot 25 \times 180°}{2196 \cdot 9}$$

$$= 40 \cdot 9°.$$

For the half-pattern draw a line 2196·9 mm in length, marking the extremities C and C'. Measure 770·1 mm from C, marking point A. Draw indefinite arcs from A and C from centre C'. Draw the second edge line from C at an angle of 40·9° (40° 54'), to determine the arc lengths.

The half-pattern can be used as a check on the dimensions obtained by the triangulation system, particularly with regard to the true length of Ba, as the frustum shown in Fig. 51 was made of sufficient taper to enable comparisons to be made with the radial-line method.

First find angle CC'B.

$$\text{Angle CC'B} = \frac{\text{Half pattern angle}}{6}$$

$$= \frac{40° 54'}{6}$$

$$= 6° 49'.$$

Next find angle C
$$= 90° - \frac{6° 49'}{2}$$

$$= 86° 35'.$$

Find length of chord AB $= 2 \sin 3° 25' \times \text{pattern radius}$
$$= 2 \times \cdot 0596 \times 2196 \cdot 9$$
$$= 261 \cdot 9 \text{ mm.}$$

Finally, find length of c.

$a = 261 \cdot 9$ mm; $b = 770 \cdot 1$ mm; C $= 86° 35'$.

$c^2 = a^2 + b^2 - 2ab \cos \text{C}$

$= 261 \cdot 9^2 + 770 \cdot 1^2 - 2 \times 261 \cdot 9 \times 770 \cdot 1 \times \cos 86° 35'$

$c = 798 \cdot 1$ mm.

Thus the length of c is shown to be 798·1 mm on the flat sheet, which is the same dimension as that obtained from the use of the triangulation formula previously given for line Ba.

OBLIQUE CONICAL FRUSTUMS

Oblique conical frustums of slight taper are generally used in air-duct work for connecting ducts of different diameter. Breeches pieces can also be made up from portions of oblique cones providing their inlets are reasonably close together.

In the following examples no cognisance has been taken of the slight curvature of connecting lines such as Cb and Dc. These lines have been treated as straight in the usual manner, as complications can arise when obtaining formulæ for the development of irregular-shaped articles if the curvature of the surface is taken into account. It is found that fairly good results can be obtained if the ordinate constants ·866 and ·5 are each reduced by ·004, thus increasing the lengths of the projector lines, with a consequent slight gain in the lengths of the pattern lines. For most purposes, however, the given formulæ are recommended for practical use when calculating true length lines for irregular-shaped objects.

The development of a half-pattern for an oblique cone frustum is shown in Fig. 53. It has a top inside diameter of 650 mm, a bottom 900 mm inside diameter, and a vertical height of 650 mm. In the plan of the job the top and base vertical centre lines are shown to be 250 mm apart. Material used is 2 mm thick.

True length lines for the pattern are found as follows :—

True length Aa
$$= \sqrt{(XY + mr - MR)^2 + h^2}$$
$$= \sqrt{(250 + 326 - 451)^2 + 650^2}$$
$$= 662 \cdot 2 \text{ mm.}$$

True length Ba
$$= \sqrt{(XY + mr - \cdot866MR)^2 + (\cdot5mr)^2 + h^2}$$
$$= \sqrt{(250 + 326 - 390\cdot6)^2 + 163^2 + 650^2}$$
$$= 695 \cdot 3 \text{ mm.}$$

True length Bb
$$= \sqrt{(XY + \cdot866mr - \cdot866MR)^2 + (\cdot5MR - \cdot5mr)^2 + h^2}$$
$$= \sqrt{(250 + 282\cdot3 - 390\cdot57)^2 + (225\cdot5 - 163)^2 + 650^2}$$
$$= 668 \cdot 2 \text{ mm.}$$

True length Cb

$$= \sqrt{(XY + \cdot866mr - \cdot5MR)^2 + (\cdot866MR - \cdot5mr)^2 + h^2}$$
$$= \sqrt{(250 + 282\cdot3 - 225\cdot5)^2 + (390\cdot57 - 163)^2 + 650^2}$$
$$= 753\cdot9 \text{ mm.}$$

True length Cc

$$= \sqrt{(XY + \cdot5mr - \cdot5MR)^2 + (\cdot866MR - \cdot866mr)^2 + h^2}$$
$$= \sqrt{(250 + 163 - 225\cdot5)^2 + (390\cdot57 - 282\cdot3)^2 + 650^2}$$
$$= 685\cdot0 \text{ mm.}$$

True length Dc

$$= \sqrt{(XY + \cdot5mr)^2 + (MR - \cdot866mr)^2 + h^2}$$
$$= \sqrt{(250 + 163)^2 + (451 - 282\cdot3)^2 + 650^2}$$
$$= 788\cdot4 \text{ mm.}$$

True length Dd

$$= \sqrt{XY^2 + (MR - mr)^2 + h^2}$$
$$= \sqrt{250^2 + (451 - 326)^2 + 650^2}$$
$$= 707\cdot5 \text{ mm.}$$

True length Ed

$$= \sqrt{(XY + \cdot5MR)^2 + (\cdot866MR - mr)^2 + h^2}$$
$$= \sqrt{(250 + 225\cdot5)^2 + (390\cdot57 - 326)^2 + h^2}$$
$$= 807\cdot9 \text{ mm.}$$

True length Ee

$$= \sqrt{(XY + \cdot5MR - \cdot5mr)^2 + (\cdot866MR - \cdot866mr)^2 + h^2}$$
$$= \sqrt{(250 + 225\cdot5 - 163)^2 + (390\cdot57 - 282\cdot3)^2 + 650^2}$$
$$= 729\cdot2 \text{ mm.}$$

True length Fe

$$= \sqrt{(XY + \cdot866MR - \cdot5mr)^2 + (\cdot866mr - \cdot5MR)^2 + h^2}$$
$$= \sqrt{(250 + 390\cdot57 - 163)^2 + (282\cdot3 - 225\cdot5)^2 + 650^2}$$
$$= 808\cdot6 \text{ mm.}$$

True length Ef

$$= \sqrt{(XY + \cdot866MR - \cdot866mr)^2 + (\cdot5MR - \cdot5mr)^2 + h^2}$$
$$= \sqrt{(250 + 390\cdot57 - 282\cdot3)^2 + (225\cdot5 - 163)^2 + 650^2}$$
$$= 744\cdot8 \text{ mm.}$$

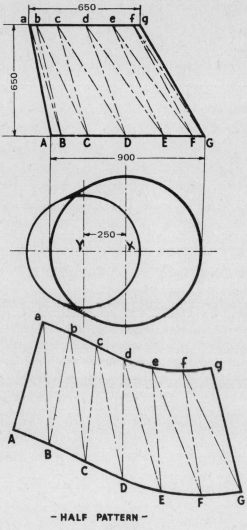

— HALF PATTERN —

Fig. 53.

True length Gf

$$= \sqrt{(XY + MR - \cdot 866mr)^2 + (\cdot 5mr)^2 + h^2}$$
$$= \sqrt{(250 + 451 - 282\cdot 3)^2 + 163^2 + 650^2}$$
$$= 790\cdot 1 \text{ mm.}$$

True length Gg

$$= \sqrt{(XY + MR - mr)^2 + h^2}$$
$$= \sqrt{(250 + 451 - 326)^2 + 650^2}$$
$$= 750\cdot 4 \text{ mm.}$$

$$\text{Length of chords AB, BC, etc.} = MR \times \cdot 522$$
$$= 451 \times \cdot 522$$
$$= 235\cdot 4 \text{ mm.}$$

$$\text{Length of chords } ab, bc, \text{ etc.} = mr \times \cdot 522$$
$$= 326 \times \cdot 522$$
$$= 170\cdot 2 \text{ mm.}$$

HALF-PATTERN FOR OBLIQUE CONICAL FRUSTUM

Draft the half-pattern in the following manner :—

First of all draw the seam Aa (662·2 mm) in a convenient position. Take Ba (695·3 mm) next, and from centre a describe an arc, and cut it in B with distance AB (235·4 mm) from centre A. Now strike an arc from centre B with true length Bb (668·2 mm). Cut this arc in b with distance ab (170·2 mm) using a as centre. Next from centre b describe an arc with true length Cb (753·9 mm), and from centre B cut this arc in C with distance BC (235·4 mm). Take true length Cc (685 mm), draw an arc from centre C and cut in c with distance bc from centre b.

Continue the rest of the development in a similar manner until the half-pattern is complete, then join the top and base points with free flowing curves.

Chapter Fifteen

TRANSFORMER AND HOPPER CALCULATIONS

THERE are many jobs, such as transformers, hoods, and hoppers, the surfaces of which are partly formed of oblique cones. Two typical examples will be given showing how formulæ for the calculation of oblique cone frustum patterns can be adapted to solve such development problems.

ROUND TO RECTANGULAR TRANSFORMER WITH SEMICIRCULAR ENDS

The first one in Fig. 54 shows the development of the half-pattern for an object which transforms from a rectangular base with semicircular ends of 300 mm radius to a circular top 400 mm diameter. Both the top and base given measurements are inside sizes. Vertical height is 250 mm, and the material used 1 mm thick.

FORMULÆ FOR TRUE LENGTH LINES

As the job is symmetrical about the vertical and horizontal centre lines of the plan, only the true lengths of one-quarter of the surface need be calculated. These lines are found as follows :—

True length Aa
$$= \sqrt{(ED + MR - mr)^2 + h^2}$$
$$= \sqrt{(150 + 300 \cdot 5 - 200 \cdot 5)^2 + 250^2}$$
$$= 353 \cdot 6 \text{ mm.}$$

True length Ba
$$= \sqrt{(ED + \cdot 866MR - mr)^2 + (\cdot 5MR)^2 + h^2}$$
$$= \sqrt{(150 + 260 \cdot 23 - 200 \cdot 5)^2 + (150 \cdot 25)^2 + 250^2}$$
$$= 359 \cdot 2 \text{ mm.}$$

True length Bb
$$= \sqrt{(ED + \cdot 866MR - \cdot 866mr)^2 + (\cdot 5MR - \cdot 5mr)^2 + h^2}$$
$$= \sqrt{(150 + 260 \cdot 23 - 173 \cdot 63)^2 + (150 \cdot 25 - 100 \cdot 25)^2 + 250^2}$$
$$= 347 \cdot 7 \text{ mm.}$$

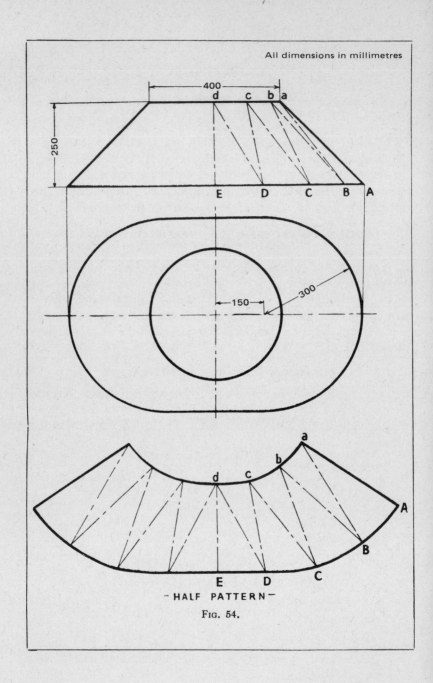

All dimensions in millimetres

400

250

d c b a

E D C B A

150   300

— HALF PATTERN —

Fig. 54.

True length Cb

$$= \sqrt{(ED + \cdot5MR - \cdot866mr)^2 + (\cdot866MR - \cdot5mr)^2 + h^2}$$
$$= \sqrt{(150 + 150\cdot25 - 173\cdot63)^2 + (260\cdot23 - 100\cdot25)^2 + 250^2}$$
$$= 322\cdot6 \text{ mm.}$$

True length Cc

$$= \sqrt{(ED + \cdot5MR - \cdot5mr)^2 + (\cdot866MR - \cdot866mr)^2 + h^2}$$
$$= \sqrt{(150 + 150\cdot25 - 100\cdot25)^2 + (260\cdot23 - 173\cdot63)^2 + 250^2}$$
$$= 331\cdot7 \text{ mm.}$$

True length Dc

$$= \sqrt{(ED - \cdot5mr)^2 + (MR - \cdot866mr)^2 + h^2}$$
$$= \sqrt{(150 - 100\cdot25)^2 + (300\cdot5 - 173\cdot63)^2 + 250^2}$$
$$= 284\cdot7 \text{ mm.}$$

True length Dd

$$= \sqrt{ED^2 + (MR - mr)^2 + h^2}$$
$$= \sqrt{150^2 + (300\cdot5 - 200\cdot5)^2 + 250^2}$$
$$= 308\cdot2 \text{ mm.}$$

True length Ed

$$= \sqrt{(MR - mr)^2 + h^2}$$
$$= \sqrt{(300\cdot5 - 200\cdot5)^2 + 250^2}$$
$$= 269\cdot3 \text{ mm.}$$

$$
\begin{aligned}
\text{Length of chords AB, BC etc.} &= MR \times \cdot522 \\
&= 300\cdot5 \times \cdot522 \\
&= 156\cdot9 \text{ mm.}
\end{aligned}
$$

$$
\begin{aligned}
\text{Length of chords } ab, bc, \text{ etc.} &= mr \times \cdot522 \\
&= 200\cdot5 \times \cdot522 \\
&= 104\cdot7 \text{ mm.}
\end{aligned}
$$

TRANSFORMER PATTERN DEVELOPMENT

To facilitate the development it is preferable to start the half-pattern, using true length Ed (269·3 mm) as a centre line, and strike off lines each side of it simultaneously. After setting down Ed, strike an arc with true length Dd (308·2 mm) from centre d, and cut it in D with distance ED (150 mm). Note that the calculation of true length Dd can be eliminated in this and similar cases by drawing ED at right angles to Ed.

Next, from centre D draw true length Dc (284·7 mm) and cut it in c with distance dc (104·7 mm). From c as centre describe an arc with distance Cc (331·7 mm). Cut this arc in C with distance DC (156·9 mm) from centre D. Continue the rest of the development in a similar manner until complete.

SEMICIRCULAR TO ROUND HOPPER WITH FLAT BACK

This component, as shown in Fig. 55, is made up of portions of an oblique cone and a rectangular to round transformer. The base of the job is 280 × 250 mm inside sizes, shaped as shown, the radius of the front being 140 mm with its centre point 40 mm from the top centre in plan. Top inside diameter is 130 mm and vertical height 140 mm. Material used, 1 mm thick. The formulæ used for obtaining the pattern true lengths are thus similar to those used for Figs. 47 and 54. Obtain each true length line as follows (bearing in mind that HL equals half the top inside diameter) :—

True length Ba'
$$= \text{vertical height} = 140 \text{ mm.}$$

True length Hb'
$$= \sqrt{(\text{HL} - \cdot 866mr)^2 + (\text{HB} - \cdot 5mr)^2 + h^2}$$
$$= \sqrt{(65 - 56\cdot72)^2 + (140 - 32\cdot75)^2 + 140^2}$$
$$= 176\cdot6 \text{ mm.}$$

True length Hc'
$$= \sqrt{(\text{HL} - \cdot 5mr)^2 + (\text{HB} - \cdot 866mr)^2 + h^2}$$
$$= \sqrt{(65 - 32\cdot75)^2 + (140 - 56\cdot72)^2 + 140^2}$$
$$= 166\cdot1 \text{ mm.}$$

True length Hd
$$= \sqrt{\text{HL}^2 + (\text{HB} - mr)^2 + h^2}$$
$$= \sqrt{65^2 + (140 - 65\cdot5)^2 + 140^2}$$
$$= 171\cdot3 \text{ mm.}$$

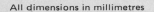
All dimensions in millimetres

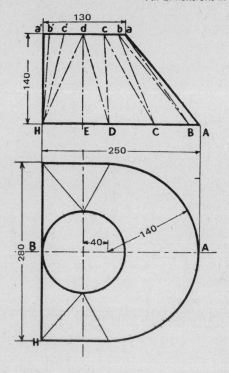

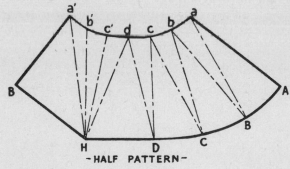

-HALF PATTERN-

Fig. 55.

True length Dd

$$= \sqrt{ED^2 + (MR - mr)^2 + h^2}$$
$$= \sqrt{40^2 + (140 \cdot 5 - 65 \cdot 5)^2 + 140^2}$$
$$= 163 \cdot 8 \text{ mm.}$$

True length Dc

$$= \sqrt{(ED - \cdot 5mr)^2 + (MR - \cdot 866mr)^2 + h^2}$$
$$= \sqrt{(40 - 32 \cdot 75)^2 + (140 \cdot 5 - 56 \cdot 72)^2 + 140^2}$$
$$= 163 \cdot 3 \text{ mm.}$$

True length Cc

$$= \sqrt{(ED + \cdot 5MR - \cdot 5mr)^2 + (\cdot 866MR - \cdot 866mr)^2 + h^2}$$
$$= \sqrt{(40 + 70 \cdot 25 - 32 \cdot 75)^2 + (121 \cdot 67 - 56 \cdot 72)^2 + 140^2}$$
$$= 172 \cdot 7 \text{ mm.}$$

True length Cb

$$= \sqrt{(ED + \cdot 5MR - \cdot 866mr)^2 + (\cdot 866MR - \cdot 5mr)^2 + h^2}$$
$$= \sqrt{(40 + 70 \cdot 25 - 56 \cdot 72)^2 + (121 \cdot 67 - 32 \cdot 75)^2 + 140^2}$$
$$= 174 \cdot 3 \text{ mm.}$$

True length Bb

$$= \sqrt{(ED + \cdot 866MR - \cdot 866mr)^2 + (\cdot 5MR - \cdot 5mr)^2 + h^2}$$
$$= \sqrt{(40 + 121 \cdot 67 - 56 \cdot 72)^2 + (70 \cdot 25 - 32 \cdot 75)^2 + 140^2}$$
$$= 178 \cdot 9 \text{ mm.}$$

True length Ba

$$= \sqrt{(ED + \cdot 866MR - mr)^2 + (\cdot 5MR)^2 + h^2}$$
$$= \sqrt{(40 + 121 \cdot 67 - 65 \cdot 5)^2 + 70 \cdot 25^2 + 140^2}$$
$$= 183 \cdot 8 \text{ mm.}$$

True length Aa

$$= \sqrt{(ED + MR - mr)^2 + h^2}$$
$$= \sqrt{(40 + 140 \cdot 5 - 65 \cdot 5)^2 + 140^2}$$
$$= 181 \cdot 2 \text{ mm.}$$

Development of the half-pattern should present no difficulties if the lay-out in Fig. 55 is closely followed. Note that there is no need to calculate the true length of Ha', as base distance HB (140 mm) is drawn at right angles to Ba' in the half-pattern.

Chapter Sixteen

BREECHES PIECE CALCULATIONS

PATTERNS for breeches pieces made up in the form of oblique cones can be readily calculated. As the branches are identical in shape and size it is only necessary to develop the full pattern for a frustum of an oblique cone and to mark off on it the correct shape of the joint curve. The complete pattern is a template for the second branch.

The development of one branch of a typical breeches piece is shown in Fig. 56. A view of the branch is depicted in Fig. 57, which pictorially presents a method of calculating the position of the joint between branches.

Inside diameter of the large end of the breeches pipe is 450 mm, and the small ends are 300 mm inside diameter, while the plan vertical centre lines of the small ends are 500 mm apart. Vertical height of the job is 600 mm, and material used 1 mm thick. True length lines for the cone frustum pattern are obtained as follows :—

True length Aa

$$= \sqrt{(XY + mr - MR)^2 + h^2}$$
$$= \sqrt{(250 + 150 \cdot 5 - 225 \cdot 5)^2 + 600^2}$$
$$= 625 \text{ mm.}$$

True length Ba

$$= \sqrt{(XY + mr - \cdot 866MR)^2 + (\cdot 5MR)^2 + h^2}$$
$$= \sqrt{(250 + 150 \cdot 5 - 195 \cdot 28)^2 + 112 \cdot 75^2 + 600^2}$$
$$= 644 \cdot 1 \text{ mm.}$$

True length Bb

$$= \sqrt{(XY + \cdot 866mr - \cdot 866MR)^2 + (\cdot 5MR - \cdot 5mr)^2 + h^2}$$
$$= \sqrt{(250 + 130 \cdot 33 - 195 \cdot 28)^2 + (112 \cdot 75 - 75 \cdot 25)^2 + 600^2}$$
$$= 629 \cdot 0 \text{ mm.}$$

131

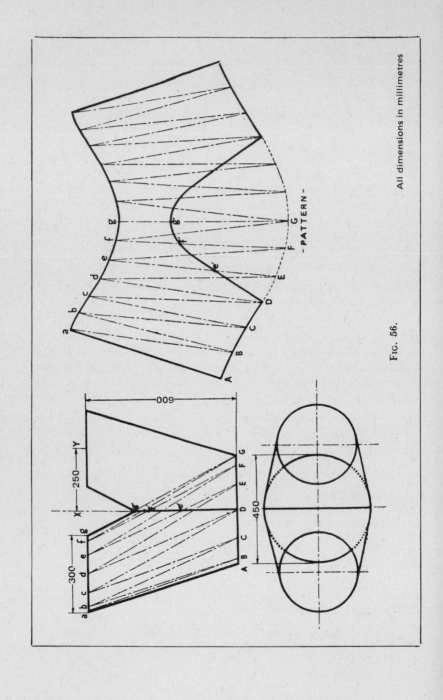

All dimensions in millimetres

FIG. 56.

True length Cb

$$= \sqrt{(XY + \cdot866mr - \cdot5MR)^2 + (\cdot866MR - \cdot5mr)^2 + h^2}$$
$$= \sqrt{(250 + 130\cdot33 - 112\cdot75)^2 + (195\cdot28 - 75\cdot25)^2 + 600^2}$$
$$= 667\cdot8 \text{ mm.}$$

True length Cc

$$= \sqrt{(XY + \cdot5mr - \cdot5MR)^2 + (\cdot866MR - \cdot866mr)^2 + h^2}$$
$$= \sqrt{(250 + 75\cdot25 - 112\cdot75)^2 + (195\cdot28 - 130\cdot33)^2 + 600^2}$$
$$= 639\cdot8 \text{ mm.}$$

True length Dc

$$= \sqrt{(XY + \cdot5mr)^2 + (MR - \cdot866mr)^2 + h^2}$$
$$= \sqrt{(250 + 75\cdot25)^2 + (225\cdot5 - 130\cdot33)^2 + 600^2}$$
$$= 689\cdot0 \text{ mm.}$$

True length Dd

$$= \sqrt{XY^2 + (MR - mr)^2 + h^2}$$
$$= \sqrt{250^2 + (225\cdot5 - 150\cdot5)^2 + 600^2}$$
$$= 654\cdot3 \text{ mm.}$$

True length Ed

$$= \sqrt{(XY + \cdot5MR)^2 + (\cdot866MR - mr)^2 + h^2}$$
$$= \sqrt{(250 + 112\cdot75)^2 + (195\cdot28 - 150\cdot5)^2 + 600^2}$$
$$= 702\cdot6 \text{ mm.}$$

True length Ee

$$= \sqrt{(XY + \cdot5MR - \cdot5mr)^2 + (\cdot866MR - \cdot866mr)^2 + h^2}$$
$$= \sqrt{(250 + 112\cdot75 - 75\cdot25)^2 + (195\cdot28 - 130\cdot33)^2 + 600^2}$$
$$= 668\cdot5 \text{ mm.}$$

True length Fe

$$= \sqrt{(XY + \cdot866MR - \cdot5mr)^2 + (\cdot866mr - \cdot5MR)^2 + h^2}$$
$$= \sqrt{(250 + 195\cdot28 - 75\cdot25)^2 + (130\cdot33 - 112\cdot75)^2 + 600^2}$$
$$= 705\cdot1 \text{ mm.}$$

True length Ff

$$= \sqrt{(XY + \cdot866MR - \cdot866mr)^2 + (\cdot5MR - \cdot5mr)^2 + h^2}$$
$$= \sqrt{(250 + 195\cdot28 - 130\cdot33)^2 + (112\cdot75 - 75\cdot25)^2 + 600^2}$$
$$= 678\cdot7 \text{ mm.}$$

True length Gf

$$= \sqrt{(XY + MR - \cdot 866mr)^2 + (\cdot 5mr)^2 + h^2}$$
$$= \sqrt{(250 + 225\cdot 5 - 130\cdot 33)^2 + 75\cdot 25^2 + 600^2}$$
$$= 696\cdot 3 \text{ mm.}$$

True length Gg

$$= \sqrt{(XY + MR - mr)^2 + h^2}$$
$$= \sqrt{(250 + 225\cdot 5 - 150\cdot 5)^2 + 600^2}$$
$$= 682\cdot 4 \text{ mm.}$$

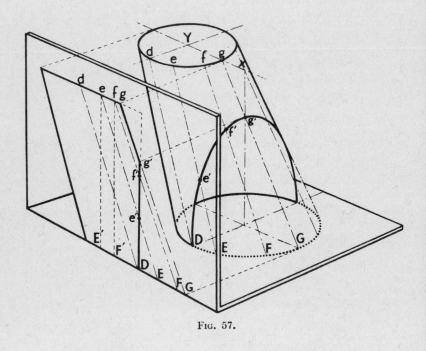

Fig. 57.

Length of chords AB, BC, etc. $= MR \times \cdot 522$
$$= 225\cdot 5 \times \cdot 522$$
$$= 117\cdot 7 \text{ mm.}$$

Length of chords ab, bc, etc. $= mr \times \cdot 522$
$$= 150\cdot 5 \times \cdot 522$$
$$= 78\cdot 6 \text{ mm.}$$

PATTERN FOR BRANCH OF BREECHES PIECE

In the given example the line Aa is made the seam, but as the branch is symmetrical about the plan horizontal centre line it is advisable to first set down true length line Gg and mark off the development lines simultaneously each side of it. Therefore, first draw Gg (682·4 mm) in a convenient position. Next, from centre G describe arcs each side of g with true length Gf (696·3 mm), and cut these arcs in f with distance fg (78·6 mm) from centre g. From centre f strike arcs with true length Ff (678·7 mm) each side of G, and cut these arcs in F from centre G with distance FG (117·7 mm). Take true length Fe (705·1 mm), describe arcs from centres F, and cut in e with distance ef from centre f. Draft the remainder of the pattern in the same manner.

Before the correct joint curve can be marked on the development, reference should be made to Fig. 57, which illustrates the correct relation of the joint cut on the surface of the object to an elevation, where it appears as a straight line Dg'. On the elevation the branch lines are shown intersecting the joint line at points g', f', and e'. The true lengths of Gg', Ff', and Ee' must be calculated to enable the correct distances to be marked off on the appropriate development lines.

FORMULÆ FOR OBTAINING JOINT LINE POINTS

Formulæ for obtaining the position of the joint cut on the branch development have been evolved, and their use is offered as a simple and accurate method of solving by calculation difficult problems appertaining to the intersection of surfaces. The true length of elevation line Ff', for instance, is found as follows :—

$$\text{True length } Ff' = \frac{\text{Base length DF}}{\text{Base length FF}'} \times \text{True length } Ff,$$

$$\text{or } \frac{\cdot 866\text{MR}}{\cdot 866\text{MR} + \text{XY} - \cdot 866 mr} \times \text{True length } Ff.$$

Calculate the joint cut true lengths as follows :—

$$\text{True length } Ee' = \frac{\cdot 5\text{MR}}{\cdot 5\text{MR} + \text{XY} - \cdot 5 mr} \times Fe$$

$$= \frac{112 \cdot 75}{112 \cdot 75 + 250 - 75 \cdot 25} \times 668 \cdot 5$$

$$= 262 \cdot 2 \text{ mm.}$$

$$\text{True length } Ff' = \frac{\cdot 866MR}{\cdot 866MR + XY - \cdot 866mr} \times Ff$$

$$= \frac{195 \cdot 28}{195 \cdot 28 + 250 - 130 \cdot 33} \times 678 \cdot 7$$

$$= 420 \cdot 8 \text{ mm.}$$

$$\text{True length } Gg' = \frac{MR}{MR + XY - mr} \times Gg$$

$$= \frac{225 \cdot 5}{225 \cdot 5 + 250 - 150 \cdot 5} \times 682 \cdot 4$$

$$= 473 \cdot 5 \text{ mm.}$$

Now measure off these distances on the branch development. From point G measure off true length Gg' (473·5 mm) on line Gg. Next, on line Ff measure point f' with true length Ff' (420·8 mm) from F, and finally measure true length Ee' (262·2 mm) from point E on line Ee. Connect the points found with a smooth curve, and join up with a straight line from point e' to D.

BREECHES PIECE WITH FLAT BACK

In the next example, shown in Fig. 58, the breeches piece is designed to lie against a flat wall or bulkhead. Each branch is 250 mm inside top diameter, the base diameter being 400 mm inside. The vertical height is 560 mm and the top centres are situated from the elevation vertical centre line a distance of 190 mm. In the plan the top and base horizontal centre lines are 75 mm apart. Material used is 1 mm thick.

As the branches are not symmetrical about any centre line, the development of the full pattern must be calculated. In this example the seam is made at the shortest portion—i.e., the throat on true length Gg.

CALCULATION OF TRUE LENGTH LINES

The true length lines for the front of the branch are obtained as follows :—

True length Aa

$$= \sqrt{(XY + mr - MR)^2 + Xy^2 + h^2}$$
$$= \sqrt{(190 + 125 \cdot 5 - 200 \cdot 5)^2 + 75^2 + 560^2}$$
$$= 576 \cdot 6 \text{ mm.}$$

True length Ba

$$= \sqrt{(XY + mr - \cdot 866MR)^2 + (XY + \cdot 5MR)^2 + h^2}$$
$$= \sqrt{(190 + 125 \cdot 5 - 173 \cdot 63)^2 + (190 + 100 \cdot 25)^2 + 560^2}$$
$$= 646 \cdot 5 \text{ mm.}$$

True length Bb

$$= \sqrt{(XY + \cdot 866mr - \cdot 866MR)^2 + (Xy + \cdot 5MR - \cdot 5mr)^2 + h^2}$$
$$= \sqrt{(190 + 108 \cdot 68 - 173 \cdot 63)^2 + (75 + 100 \cdot 25 - 62 \cdot 75)^2 + 560^2}$$
$$= 584 \cdot 7 \text{ mm.}$$

True length Cb

$$= \sqrt{(XY + \cdot 866mr - \cdot 5MR)^2 + (Xy + \cdot 866MR - \cdot 5mr)^2 + h^2}$$
$$= \sqrt{(190 + 108 \cdot 68 - 100 \cdot 25)^2 + (75 + 173 \cdot 63 - 62 \cdot 75)^2 + 560^2}$$
$$= 622 \cdot 5 \text{ mm.}$$

True length Cc

$$= \sqrt{(XY + \cdot 5mr - \cdot 5MR)^2 + (Xy + \cdot 866MR - \cdot 866mr)^2 + h^2}$$
$$= \sqrt{(190 + 62 \cdot 75 - 100 \cdot 25)^2 + (75 + 173 \cdot 63 - 108 \cdot 68)^2 + 560^2}$$
$$= 597 \cdot 0 \text{ mm.}$$

True length Dc

$$= \sqrt{(XY + \cdot 5mr)^2 + (Xy + MR - \cdot 866mr)^2 + h^2}$$
$$= \sqrt{(190 + 62 \cdot 75)^2 + (75 + 173 \cdot 63 - 108 \cdot 68)^2 + 560^2}$$
$$= 630 \cdot 1 \text{ mm.}$$

True length Dd

$$= \sqrt{XY^2 + (Xy + MR - mr)^2 + h^2}$$
$$= \sqrt{190^2 + (75 + 200 \cdot 5 - 125 \cdot 5)^2 + 560^2}$$
$$= 610 \cdot 1 \text{ mm.}$$

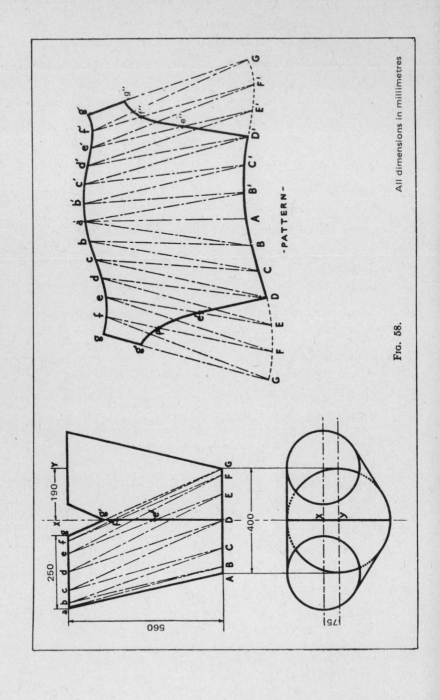

FIG. 58.

-PATTERN-

All dimensions in millimetres

True length Ed

$$= \sqrt{(XY + \cdot 5MR)^2 + (Xy + \cdot 866MR - mr)^2 + h^2}$$
$$= \sqrt{(190 + 100 \cdot 25)^2 + (75 + 173 \cdot 63 - 125 \cdot 5)^2 + 560^2}$$
$$= 642 \cdot 6 \text{ mm.}$$

True length Ee

$$= \sqrt{(XY + \cdot 5MR - \cdot 5mr)^2 + (Xy + \cdot 866MR - \cdot 866mr)^2 + h^2}$$
$$= \sqrt{(190 + 100 \cdot 25 - 62 \cdot 75)^2 + (75 + 173 \cdot 63 - 108 \cdot 68)^2 + 560^2}$$
$$= 620 \cdot 4 \text{ mm.}$$

True length Fe

$$= \sqrt{(XY + \cdot 866MR - \cdot 5mr)^2 + (Xy + \cdot 5MR - \cdot 866mr)^2 + h^2}$$
$$= \sqrt{(190 + 173 \cdot 63 - 62 \cdot 75)^2 + (75 + 100 \cdot 25 - 108 \cdot 68)^2 + 560^2}$$
$$= 639 \cdot 2 \text{ mm.}$$

True length Ff

$$= \sqrt{(XY + \cdot 866MR - \cdot 866mr)^2 + (Xy + \cdot 5MR - \cdot 5mr)^2 + h^2}$$
$$= \sqrt{(190 + 173 \cdot 63 - 108 \cdot 68)^2 + (75 + 100 \cdot 25 - 62 \cdot 75)^2 + 560^2}$$
$$= 625 \cdot 5 \text{ mm.}$$

True length Gf

$$= \sqrt{(XY + MR - \cdot 866mr)^2 + (Xy + \cdot 5mr)^2 + h^2}$$
$$= \sqrt{(190 + 200 \cdot 5 - 108 \cdot 68)^2 + (75 + 62 \cdot 75)^2 + 560^2}$$
$$= 641 \cdot 9 \text{ mm.}$$

True length Gg

$$= \sqrt{(XY + MR - mr)^2 + Xy^2 + h^2}$$
$$= \sqrt{(190 + 200 \cdot 5 - 125 \cdot 5)^2 + 75^2 + 560^2}$$
$$= 624 \cdot 1 \text{ mm.}$$

Next calculate the true lengths for the back portion of the branch :—

True length B$'a$

$$= \sqrt{(XY + mr - \cdot 866MR)^2 + (\cdot 5MR - Xy)^2 + h^2}$$
$$= \sqrt{(190 + 125 \cdot 5 - 173 \cdot 63)^2 + (100 \cdot 25 - 75)^2 + 560^2}$$
$$= 578 \cdot 2 \text{ mm.}$$

True length B$'b'$

$$= \sqrt{(XY + \cdot 866mr - \cdot 866MR)^2 + (Xy + \cdot 5mr - \cdot 5MR)^2 + h^2}$$

$$= \sqrt{(190 + 108 \cdot 68 - 173 \cdot 63)^2 + (75 + 62 \cdot 75 - 100 \cdot 25)^2 + 560^2}$$
$$= 575 \cdot 0 \text{ mm.}$$

True length C'b'

$$= \sqrt{(XY + \cdot 866mr - \cdot 5MR)^2 + (Xy + \cdot 5mr - \cdot 866MR)^2 + h^2}$$
$$= \sqrt{(190 + 108 \cdot 68 - 100 \cdot 25)^2 + (75 + 62 \cdot 75 - 173 \cdot 63)^2 + 560^2}$$
$$= 595 \cdot 2 \text{ mm.}$$

True length C'c'

$$= \sqrt{(XY + \cdot 5mr - \cdot 5MR)^2 + (Xy + \cdot 866mr - \cdot 866MR)^2 + h^2}$$
$$= \sqrt{(190 + 62 \cdot 75 - 100 \cdot 25)^2 + (75 + 108 \cdot 68 - 173 \cdot 63)^2 + 560^2}$$
$$= 580 \cdot 5 \text{ mm.}$$

True length D'c'

$$= \sqrt{(XY + \cdot 5mr)^2 + (Xy + \cdot 866mr - MR)^2 + h^2}$$
$$= \sqrt{(190 + 62 \cdot 75)^2 + (75 + 108 \cdot 63 - 200 \cdot 5)^2 + 560^2}$$
$$= 614 \cdot 5 \text{ mm.}$$

True length D'd'

$$= \sqrt{XY^2 + (Xy + mr - MR)^2 + h^2}$$
$$= \sqrt{190^2 + (75 + 125 \cdot 5 - 200 \cdot 5)^2 + 560^2}$$
$$= 591 \cdot 4 \text{ mm.}$$

True length E'd'

$$= \sqrt{(XY + \cdot 5MR)^2 + (Xy + mr - \cdot 866MR)^2 + h^2}$$
$$= \sqrt{(190 + 100 \cdot 25)^2 + (75 + 125 \cdot 5 - 173 \cdot 63)^2 + 560^2}$$
$$= 631 \cdot 3 \text{ mm.}$$

True length E'e'

$$= \sqrt{(XY + \cdot 5MR - \cdot 5mr)^2 + (Xy + \cdot 866mr - \cdot 866MR)^2 + h^2}$$
$$= \sqrt{(190 + 100 \cdot 25 - 62 \cdot 75)^2 + (75 + 108 \cdot 68 - 173 \cdot 63)^2 + 560^2}$$
$$= 604 \cdot 5 \text{ mm.}$$

True length F'e'

$$= \sqrt{(XY + \cdot 866MR - \cdot 5mr)^2 + (Xy + \cdot 866mr - \cdot 5MR)^2 + h^2}$$
$$= \sqrt{(190 + 173 \cdot 63 - 62 \cdot 75)^2 + (75 + 108 \cdot 68 - 100 \cdot 25)^2 + 560^2}$$
$$= 641 \cdot 0 \text{ mm.}$$

True length F'f'

$$= \sqrt{(XY + \cdot866MR - \cdot866mr)^2 + (Xy + \cdot5mr - \cdot5MR)^2 + h^2}$$
$$= \sqrt{(190 + 173\cdot63 - 108\cdot68)^2 + (75 + 62\cdot75 - 100\cdot25)^2 + 560^2}$$
$$= 616\cdot4 \text{ mm.}$$

True length Gf'

$$= \sqrt{(XY + MR - \cdot866mr)^2 + (Xy - \cdot5mr)^2 + h^2}$$
$$= \sqrt{(190 + 200\cdot5 - 108\cdot68)^2 + (75 - 62\cdot75)^2 + 560^2}$$
$$= 627\cdot0 \text{ mm.}$$

$$\begin{aligned}
\text{Length of chords AB, BC, etc.} &= MR \times \cdot522 \\
&= 200\cdot5 \times \cdot522 \\
&= 104\cdot7 \text{ mm.}
\end{aligned}$$

$$\begin{aligned}
\text{Length of chords } ab,\ bc,\ \text{etc.} &= mr \times \cdot522 \\
&= 125\cdot5 \times \cdot522 \\
&= 65\cdot5 \text{ mm.}
\end{aligned}$$

DEVELOPMENT OF THE BRANCH PATTERN

Start the development by drawing in a convenient position true length Gg (624·1 mm). From point G mark off true length Gf (641·9 mm) from centre G, and cut this in f with distance fg (65·5 mm) marked off from centre g. Take true length Ff (625·5 mm), and from centre f describe an arc and cut it in F from centre G with distance FG (104·7 mm). Continue the laying-out of the development similarly until the drafting of the cone frustum is complete, then calculate the position of the joint points as follows :—

$$\begin{aligned}
\text{True length E}e' &= \frac{\cdot5MR}{\cdot5MR + XY - \cdot5mr} \times Ee \\
&= \frac{100\cdot25}{100\cdot25 + 190 - 62\cdot75} \times 602\cdot4 \\
&= 265\cdot5 \text{ mm.}
\end{aligned}$$

$$\begin{aligned}
\text{True length F}f' &= \frac{\cdot866MR}{\cdot866MR + XY - \cdot866mr} \times Ff \\
&= \frac{173\cdot63}{173\cdot63 + 190 - 108\cdot68} \times 625\cdot5 \\
&= 426\cdot0 \text{ mm.}
\end{aligned}$$

$$\text{True length } Gg' = \frac{MR}{MR + XY - mr} \times Gg$$

$$= \frac{200 \cdot 5}{200 \cdot 5 + 190 - 125 \cdot 5} \times 624 \cdot 1$$

$$472 \cdot 1 \text{ mm.}$$

$$\text{True length } F'f'' = \frac{\cdot 866 MR}{\cdot 866 MR + XY - \cdot 866 mr} \times F'f'$$

$$= \frac{173 \cdot 63}{173 \cdot 63 + 190 - 108 \cdot 68} \times 616 \cdot 4$$

$$= 419 \cdot 8 \text{ mm.}$$

$$\text{True length } E'e'' = \frac{\cdot 5 MR}{\cdot 5 MR + XY - \cdot 5 mr} \times E'e'$$

$$= \frac{100 \cdot 25}{100 \cdot 25 + 190 - 62 \cdot 75} \times 604 \cdot 5$$

$$= 266 \cdot 5 \text{ mm.}$$

After obtaining these distances, mark off on the appropriate lines as shown. On the pattern end lines Gg measure off from G true length Gg' (472·1 mm) marking g'. Next on pattern line Ff measure off from F true length Ff' (426 mm) to obtain point f'. Similarly find point f'' on $F'f'$ by measuring true length $F'f''$ (419·8 mm) from F'. From point E on Ee mark off e' with distance Ee' (265·5 mm). Finally obtain point e'' (266·5 mm) from point E'. Join up all the points as shown in the diagram to complete the pattern.

Chapter Seventeen

JUNCTION PIECE CALCULATIONS

I N all cases where it is possible to obtain the patterns of some of the more complicated types of multiple-way junction pieces by calculation, it will be found that the preparatory work necessary when using geometrical lay-out methods is eliminated. Much of the difficulty in developing work of this nature, as a rule, is to set out the correct position of the joint line, yet this can be easily calculated by an adaptation of formulæ previously given.

The three-way piece shown in Fig. 59 is drawn with one of the inlets situated on the plan horizontal centre line, as this is the usual position in which the job is set-out in the workshop for development purposes. Using calculation methods, however, it is immaterial in which position the plan or elevation of the multiple-way piece is drawn on the working drawing, as all the information that is required for the pattern developments are the dimensions plus suitable formulæ.

In the given example the base dimension of the three-way piece is 450 mm inside diameter, and the three small ends are each 250 mm inside diameter. The vertical height of the job is 600 mm, and the material used is 1 mm thick.

Calculate the pattern true lengths as follows :—

True length Aa

$$= \sqrt{(XY + mr - MR)^2 + h^2}$$
$$= \sqrt{(225 + 125 \cdot 5 - 225 \cdot 5)^2 + 600^2}$$
$$= 612 \cdot 8 \text{ mm.}$$

True length Ba

$$= \sqrt{(XY + mr - \cdot 886MR)^2 + (\cdot 5MR)^2 + h^2}$$
$$= \sqrt{(225 + 125 \cdot 5 - 195 \cdot 28)^2 + 112 \cdot 75^2 + 600^2}$$
$$= 629 \cdot 9 \text{ mm.}$$

True length Bb

$$= \sqrt{(XY + \cdot866mr - \cdot866MR)^2 + (\cdot5MR - \cdot5mr)^2 + h^2}$$
$$= \sqrt{(225 + 108\cdot68 - 195\cdot28)^2 + (112\cdot75 - 62\cdot75)^2 + 600^2}$$
$$= 617\cdot8 \text{ mm.}$$

True length Cb

$$= \sqrt{(XY + \cdot886mr - \cdot5MR)^2 + (\cdot866MR - \cdot5mr)^2 + h^2}$$
$$= \sqrt{(225 + 108\cdot68 - 112\cdot75)^2 + (195\cdot28 - 62\cdot75)^2 + 600^2}$$
$$= 653\cdot0 \text{ mm.}$$

True length Cc

$$= \sqrt{(XY + \cdot5mr - \cdot5MR)^2 + (\cdot866MR - \cdot866mr)^2 + h^2}$$
$$= \sqrt{(225 + 62\cdot75 - 112\cdot75)^2 + (195\cdot28 - 108\cdot68)^2 + 600^2}$$
$$= 630\cdot9 \text{ mm.}$$

True length Dc

$$= \sqrt{(XY + \cdot5mr)^2 + (MR - \cdot866mr)^2 + h^2}$$
$$= \sqrt{(225 + 62\cdot75)^2 + (225\cdot5 - 108\cdot68)^2 + 600^2}$$
$$= 675\cdot6 \text{ mm.}$$

True length Dd

$$= \sqrt{XY^2 + (MR - mr)^2 + h^2}$$
$$= \sqrt{225^2 + (225\cdot5 - 125\cdot5)^2 + 600^2}$$
$$= 648\cdot6 \text{ mm.}$$

True length Ed

$$= \sqrt{(XY + \cdot5MR)^2 + (\cdot866MR - mr)^2 + h^2}$$
$$= \sqrt{(225 + 112\cdot75)^2 + (195\cdot28 - 125\cdot5)^2 + 600^2}$$
$$= 692\cdot1 \text{ mm.}$$

True length Ee

$$= \sqrt{(XY + \cdot5MR - \cdot5mr)^2 + (\cdot866MR - \cdot866mr)^2 + h^2}$$
$$= \sqrt{(225 + 112\cdot75 - 62\cdot75)^2 + (195\cdot28 - 108\cdot68)^2 + 600^2}$$
$$= 665\cdot7 \text{ mm.}$$

True length Fe

$$= \sqrt{(XY + \cdot866MR - \cdot5mr)^2 + (\cdot5MR - \cdot866mr)^2 + h^2}$$
$$= \sqrt{(225 + 195\cdot28 - 62\cdot75)^2 + (112\cdot75 - 108\cdot68)^2 + 600^2}$$
$$= 698\cdot4 \text{ mm.}$$

True length Ff

$$= \sqrt{(XY + \cdot866MR - \cdot866mr)^2 + (\cdot5MR - \cdot5mr)^2 + h^2}$$
$$= \sqrt{(225 + 195\cdot28 - 108\cdot68)^2 + (112\cdot75 - 62\cdot75)^2 + 600^2}$$
$$= 677\cdot9 \text{ mm.}$$

True length Gf

$$= \sqrt{(XY + MR - \cdot866mr)^2 + (\cdot5mr)^2 + h^2}$$
$$= \sqrt{(225 + 225\cdot5 - 108\cdot68)^2 + 62\cdot75^2 + 600^2}$$
$$= 693\cdot4 \text{ mm.}$$

True length Gg

$$= \sqrt{(XY + MR - mr)^2 + h^2}$$
$$= \sqrt{(225 + 225\cdot5 - 125\cdot5)^2 + 600^2}$$
$$= 682\cdot4 \text{ mm.}$$

$$\text{Lengths of chords AB, BC, etc.} = MR \times \cdot522$$
$$= 225\cdot5 \times \cdot522$$
$$= 117\cdot7 \text{ mm.}$$
$$\text{Lengths of chords } ab, bc, \text{ etc.} = mr \times \cdot522$$
$$= 125\cdot5 \times \cdot522$$
$$= 65\cdot5 \text{ mm.}$$

DEVELOPMENT OF THREE-WAY PATTERN

To commence the pattern lay-out, draw in a suitable position true length line Aa (612·8 mm), and construct true length lines Ba, Bb, Cb, etc., on each side of this line as follows :—

Take true length Ba (629·9 mm), and from a as centre strike an arc. Cut this arc in B with distance AB (117·7 mm), using A as centre. Take true length Bb (617·8 mm), and from centre B describe an arc, cutting it in b with distance ab (65·5 mm) from centre a. Next, from centre b strike true length Cb (653 mm), and cut the arc in C with distance BC. Continue on similar lines until the pattern lay-out for the full cone frustum is complete.

CALCULATION OF POSITION OF JOINT LINE

The formulæ for obtaining the position of the joint line points are adapted from those which have already been given

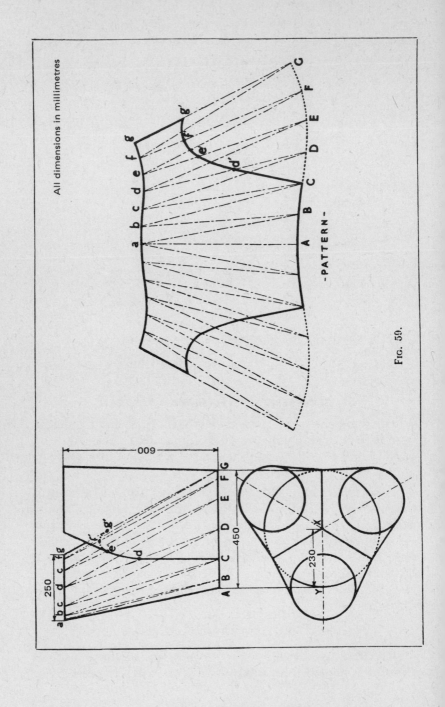

All dimensions in millimetres

-PATTERN-

Fig. 59.

for the breeches piece.   The lengths of the lines are obtained as follows :—

$$\text{True length } Dd' = \frac{\cdot 5MR}{\cdot 5MR + \cdot 866XY - \cdot 5mr} \times Dd$$

$$= \frac{112 \cdot 75}{112 \cdot 75 + 194 \cdot 85 - 62 \cdot 75} \times 648 \cdot 6$$

$$= 298 \cdot 7 \text{ mm.}$$

$$\text{True length } Ee' = \frac{\cdot 866MR}{\cdot 866MR + \cdot 866XY - \cdot 866mr} \times Ee$$

$$= \frac{195 \cdot 28}{195 \cdot 28 + 194 \cdot 85 - 108 \cdot 68} \times 665 \cdot 7$$

$$= 461 \cdot 9 \text{ mm.}$$

$$\text{True length } Ff' = \frac{MR}{MR + \cdot 866XY - mr} \times Ff$$

$$= \frac{225 \cdot 5}{225 \cdot 5 + 194 \cdot 85 - 125 \cdot 5} \times 677 \cdot 9$$

$$= 518 \cdot 5 \text{ mm.}$$

$$\text{True length } Gg' = \frac{MR}{MR + XY - mr} \times Gg$$

$$= \frac{225 \cdot 5}{225 \cdot 5 + 225 - 125 \cdot 5} \times 682 \cdot 4$$

$$= 473 \cdot 5 \text{ mm.}$$

Measure off from G on line Gg in the pattern lay-out, true length Gg' (473·5 mm).   Take true length Ff' (518·5 mm) and mark off point f' from point F on line Ff.   Next, mark off from E true length Ee' (461·9 mm) on pattern line Ee.   Finally measure off true length Dd' (298·7 mm) from point D on line Dd.   Connect g', f', e', and d' with a smooth curve, and join d' to C.   This completes the template for one branch, from which two more branches must be marked to complete the laying-out of the job.

CONNECTING PIECE BETWEEN OBLIQUE PLANES

With the exception of the job shown in Fig. 50 all the examples explained have had their top and bottom surfaces lying between parallel planes.   For most jobs which have end surfaces lying obliquely to each other it is a slightly more difficult process to obtain the true length lines for the pattern development, because the elevation triangles are of different heights. Therefore the correct heights of the elevation lines must be calculated before the true length pattern lines are obtainable.

The example in Fig. 60 is of a connecting piece with a top 200 mm inside diameter, which is inclined at an angle of 25° to a base 330 mm inside diameter. The vertical height at the centre line is 180 mm, and the material used is 2 mm thick. Before calculating the true length of the pattern lines it is necessary to obtain the heights ha, hb, hc, etc., and also the length of half the minor axis (plan radius, or pr) of an ellipse, which is the projection of the inclined top in the elevation.

The heights of the vertical lines ha, hb, and hc are found by multiplying the sine of the top angle by the mean radius, ·866 mean radius, and ·5 mean radius, respectively, and *subtracting* the figures obtained from the vertical height hd (180 mm). Heights of the vertical lines he, hf, and hg are obtained by multiplying the sine of the top angle by ·5 mean radius, ·866 mean radius, and mean radius, respectively, and *adding* these to the vertical height.

The various heights for the given example are found as follows :—

$$
\begin{aligned}
\text{Vertical height } ha &= hd - (\sin 25° \times mr) \\
&= 180 - (·4226 \times 101) \\
&= 137·3 \text{ mm.}
\end{aligned}
$$

$$
\begin{aligned}
\text{Vertical height } hb &= hd - (\sin 25° \times ·866mr) \\
&= 180 - (·4226 \times 87·47) \\
&= 143·0 \text{ mm.}
\end{aligned}
$$

$$
\begin{aligned}
\text{Vertical height } hc &= hd - (\sin 25° \times ·5mr) \\
&= 180 - (·4226 \times 50·5) \\
&= 158·7 \text{ mm.}
\end{aligned}
$$

$$
\begin{aligned}
\text{Vertical height } he &= hd + (\sin 25° \times ·5mr) \\
&= 180 + (·4226 \times 50·5) \\
&= 201·3 \text{ mm.}
\end{aligned}
$$

$$
\begin{aligned}
\text{Vertical height } hf &= hd + (\sin 25° \times ·866mr) \\
&= 180 + (·4226 \times 87·47) \\
&= 217·0 \text{ mm.}
\end{aligned}
$$

Vertical height $hg = hd + (\sin 25° \times mr)$
$$= 180 + (\cdot4226 \times 101)$$
$$= 222\cdot7 \text{ mm.}$$

The plan radius is obtained by multiplying the mean radius by the cosine of the angle. In the given example, therefore :—

$$pr = mr \times \cos 25°$$
$$= 101 \times \cdot9063$$
$$= 91\cdot54 \text{ mm.}$$

Obtain the true length lines as follows :—

True length Aa

$$= \sqrt{(\text{MR} - pr)^2 + ha^2}$$
$$= \sqrt{(166 - 91\cdot54)^2 + 137\cdot3^2}$$
$$= 156\cdot2 \text{ mm.}$$

True length Ba

$$= \sqrt{(\cdot866\text{MR} - pr)^2 + (\cdot5\text{MR})^2 + ha^2}$$
$$= \sqrt{(143\cdot76 - 91\cdot54)^2 + (83)^2 + 137\cdot3^2}$$
$$= 168\cdot7 \text{ mm.}$$

True length Bb

$$= \sqrt{(\cdot866\text{MR} - \cdot866pr)^2 + (\cdot5\text{MR} - \cdot5mr)^2 + hb^2}$$
$$= \sqrt{(143\cdot76 - 79\cdot27)^2 + (83 - 50\cdot5)^2 + 143^2}$$
$$= 160\cdot2 \text{ mm.}$$

True length Cb

$$= \sqrt{(\cdot866pr - \cdot5\text{MR})^2 + (\cdot866\text{MR} - \cdot5mr)^2 + hb^2}$$
$$= \sqrt{(79\cdot27 - 83)^2 + (143\cdot76 - 50\cdot5)^2 + 143^2}$$
$$= 170\cdot8 \text{ mm.}$$

True length Cc

$$= \sqrt{(\cdot5\text{MR} - \cdot5pr)^2 + (\cdot866\text{MR} - \cdot866mr)^2 + hc^2}$$
$$= \sqrt{(83 - 45\cdot77)^2 + (143\cdot76 - 87\cdot47)^2 + 158\cdot7^2}$$
$$= 269\cdot9 \text{ mm.}$$

True length Dc

$$= \sqrt{(\cdot5pr)^2 + (\text{MR} - \cdot866mr)^2 + hc^2}$$
$$= \sqrt{45\cdot77^2 + (166 - 87\cdot47)^2 + 158\cdot7^2}$$
$$= 182\cdot9 \text{ mm.}$$

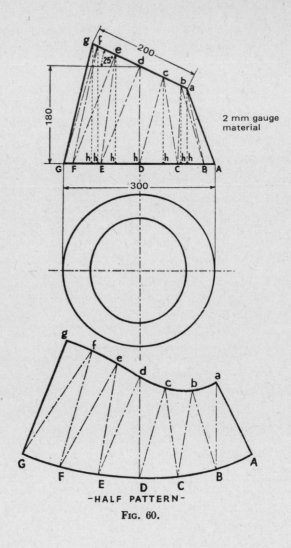

All dimensions in millimetres

2 mm gauge material

-HALF PATTERN-

Fig. 60.

True length Dd

$$= \sqrt{(MR - mr)^2 + hd^2}$$
$$= \sqrt{(166 - 101)^2 + 180^2}$$
$$= 191 \cdot 4 \text{ mm.}$$

True length Ed

$$= \sqrt{(\cdot 5MR)^2 + (\cdot 866MR - mr)^2 + hd^2}$$
$$= \sqrt{83^2 + (143 \cdot 76 - 101)^2 + 180^2}$$
$$= 202 \cdot 8 \text{ mm.}$$

True length Ee

$$= \sqrt{(\cdot 5MR - \cdot 5pr)^2 + (\cdot 866MR - \cdot 866mr)^2 + he^2}$$
$$= \sqrt{(83 - 45 \cdot 77)^2 + (143 \cdot 76 - 87 \cdot 47)^2 + 201 \cdot 3^2}$$
$$= 212 \cdot 3 \text{ mm.}$$

True length Fe

$$= \sqrt{(\cdot 866MR - \cdot 5pr)^2 + (\cdot 866mr - \cdot 5MR)^2 - he^2}$$
$$= \sqrt{(143 \cdot 76 - 45 \cdot 77)^2 + (87 \cdot 47 - 83)^2 + 201 \cdot 3^2}$$
$$= 228 \cdot 3 \text{ mm.}$$

True length Ff

$$= \sqrt{(\cdot 866MR - \cdot 866pr)^2 + (\cdot 5MR - \cdot 5mr)^2 + hf^2}$$
$$= \sqrt{(143 \cdot 76 - 79 \cdot 27)^2 + (83 - 50 \cdot 5)^2 + 217^2}$$
$$= 228 \cdot 7 \text{ mm.}$$

True length Gf

$$= \sqrt{(MR - \cdot 866pr)^2 + (\cdot 5mr)^2 + hf^2}$$
$$= \sqrt{(166 - 79 \cdot 27)^2 + 50 \cdot 5^2 + 217^2}$$
$$= 239 \cdot 1 \text{ mm.}$$

True length Gg

$$= \sqrt{(MR - pr)^2 + hg^2}$$
$$= \sqrt{(166 - 91 \cdot 54)^2 + 222 \cdot 7^2}$$
$$= 234 \cdot 9 \text{ mm.}$$

$$\text{Length of chords AB, BC, etc.} = MR \times \cdot 522$$
$$= 166 \times \cdot 522$$
$$= 86 \cdot 7 \text{ mm.}$$

Length of chords ab, bc, etc. $= mr \times \cdot522$
$$= 101 \times \cdot522$$
$$= 52\cdot7 \text{ mm.}$$

DRAFTING THE PATTERN

As the job is symmetrical about the horizontal plan centre line, it is only necessary to develop a half-pattern, the remainder of the job, of course, being a repetition of the lines found. Start the pattern with the seam Aa (156·2 mm), and from centre a describe an arc with true length Ba (168·7 mm). Cut this arc in B from centre A with distance AB (86·7 mm). Take true length Bb (160·2 mm), strike an arc from B, and cut in b with ab (52·7 mm). Next, from centre b describe an arc with true length Cb (170·8 mm), and cut in C with distance BC.

The remainder of the pattern should be constructed in similar fashion, as shown in the diagram.

TWO-WAY BREECHES PIECE

The two-way breeches piece shown in Fig. 61 has its circular inlets inclined at an angle of 30° to a base of similar shape, therefore the branches are not geometrically portions of oblique cones. This fact, however, does not affect the principle of pattern calculation for the branches, as the formulæ are similar to those previously given.

Base inside diameter of the breeches piece is 1000 mm, the top inside diameters 700 mm, and the vertical height to the top centre 1000 mm. Material used is 1 mm gauge. The branches are off-centre 580 mm, as at XY in the plan.

CALCULATION OF PATTERN LINES

First calculate vertical heights ha, hb, hc, etc., as follows :—

Vertical height $ha = hd - (\sin 30° \times mr)$
$$= 1000 - (\cdot5 \times 350\cdot5)$$
$$= 824\cdot8 \text{ mm.}$$
Vertical height $hb = hd - (\sin 30° \times \cdot866mr)$
$$= 1000 - (\cdot5 \times 303\cdot5)$$
$$= 848\cdot2 \text{ mm.}$$

Vertical height $hc = hd - (\sin 30° \times \cdot5mr)$
$$= 1000 - (\cdot5 \times 175 \cdot 25)$$
$$= 912 \cdot 4 \text{ mm.}$$

Vertical height $he = hd + (\sin 30° \times \cdot5mr)$
$$= 1000 + (\cdot5 \times 175 \cdot 25)$$
$$= 1087 \cdot 6 \text{ mm.}$$

Vertical height $hf = hd + (\sin 30° \times \cdot866mr)$
$$= 1000 + (\cdot5 \times 303 \cdot 5)$$
$$= 1151 \cdot 8 \text{ mm.}$$

Vertical height $hg = hd + (\sin 30° \times mr)$
$$= 1000 + (\cdot5 \times 350 \cdot 5)$$
$$= 1175 \cdot 3 \text{ mm.}$$

Next obtain plan radius (pr)
$$pr = mr \times \cos 30°$$
$$= 350 \cdot 5 \times \cdot866$$
$$= 303 \cdot 53.$$

Now find true length pattern lines :—

True length Aa
$$= \sqrt{(XY + pr - MR)^2 + ha^2}$$
$$= \sqrt{(580 + 303 \cdot 53 - 500 \cdot 5)^2 + 824 \cdot 8^2}$$
$$= 909 \cdot 4 \text{ mm.}$$

True length Ba
$$= \sqrt{(XY + pr - \cdot866MR)^2 + (\cdot5MR)^2 + ha^2}$$
$$= \sqrt{(580 + 303 \cdot 53 - 433 \cdot 43)^2 + 250 \cdot 25^2 + 824 \cdot 8^2}$$
$$= 977 \cdot 4 \text{ mm.}$$

True length Bb
$$= \sqrt{(XY + \cdot866pr - \cdot866MR)^2 + (\cdot5MR - \cdot5mr)^2 + hb^2}$$
$$= \sqrt{(580 + 262 \cdot 85 - 433 \cdot 43)^2 + (250 \cdot 25 - 175 \cdot 25)^2 + 848 \cdot 2^2}$$
$$= 944 \cdot 6 \text{ mm.}$$

True length Cb
$$= \sqrt{(XY + \cdot866pr - \cdot5MR)^2 + (\cdot866MR - \cdot5mr)^2 + hb^2}$$
$$= \sqrt{(580 + 262 \cdot 85 - 250 \cdot 25)^2 + (433 \cdot 43 - 175 \cdot 25)^2 + 848 \cdot 2^2}$$
$$= 1066 \cdot 5 \text{ mm.}$$

True length C*c*

$$= \sqrt{(XY + \cdot 5pr - \cdot 5MR)^2 + (\cdot 866MR - \cdot 866mr)^2 + hc^2}$$
$$= \sqrt{(580 + 151\cdot 76 - 250\cdot 25)^2 + (433\cdot 43 - 303\cdot 5)^2 + 912\cdot 4^2}$$
$$= 1039\cdot 4 \text{ mm.}$$

True length D*c*

$$= \sqrt{(XY + \cdot 5pr)^2 + (MR - \cdot 866mr)^2 + hc^2}$$
$$= \sqrt{(580 + 151\cdot 76)^2 + (500\cdot 5 - 303\cdot 5)^2 + 912\cdot 4^2}$$
$$= 1185\cdot 6 \text{ mm.}$$

True length D*d*

$$= \sqrt{XY^2 + (MR - mr)^2 + hd^2}$$
$$= \sqrt{580^2 + (500\cdot 5 - 350\cdot 5)^2 + 1000^2}$$
$$= 1165\cdot 7 \text{ mm.}$$

True length E*d*

$$= \sqrt{(XY + \cdot 5MR)^2 + (\cdot 866MR - mr)^2 + hd^2}$$
$$= \sqrt{(580 + 250\cdot 25)^2 + (433\cdot 43 - 350\cdot 5)^2 + 1000^2}$$
$$= 1325\cdot 9 \text{ mm.}$$

True length E*e*

$$= \sqrt{(XY + \cdot 5MR - \cdot 5pr)^2 + (\cdot 866MR - \cdot 866mr)^2 + he^2}$$
$$= \sqrt{(580 + 250\cdot 25 - 151\cdot 76)^2 + (433\cdot 43 - 303\cdot 5)^2 + 1087\cdot 6^2}$$
$$= 1288\cdot 0 \text{ mm.}$$

True length F*e*

$$= \sqrt{(XY + \cdot 866MR - \cdot 5pr)^2 + (\cdot 866mr - \cdot 5MR)^2 + he^2}$$
$$= \sqrt{(580 + 433\cdot 43 - 151\cdot 76)^2 + (303\cdot 5 - 250\cdot 25)^2 + 1087\cdot 6^2}$$
$$= 1388\cdot 4 \text{ mm.}$$

True length F*f*

$$= \sqrt{(XY + \cdot 866MR - \cdot 866pr)^2 + (\cdot 5MR - \cdot 5mr)^2 + hf^2}$$
$$= \sqrt{(580 + 433\cdot 43 - 262\cdot 86)^2 + (250\cdot 25 - 175\cdot 25)^2 + 1151\cdot 8^2}$$
$$= 1376\cdot 7 \text{ mm.}$$

True length G*f*

$$= \sqrt{(XY + MR - \cdot 866mr)^2 + (\cdot 5mr)^2 + hf^2}$$
$$= \sqrt{(580 + 500\cdot 25 - 262\cdot 86)^2 + 175\cdot 25^2 + 1151\cdot 8^2}$$
$$= 1423\cdot 0 \text{ mm.}$$

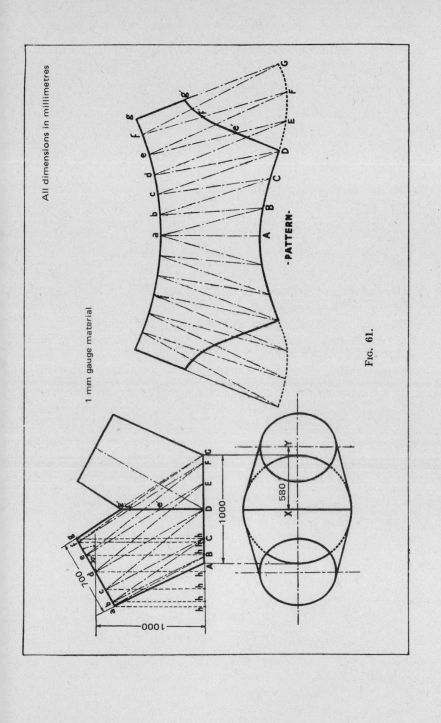

All dimensions in millimetres

1 mm gauge material

- PATTERN -

Fig. 61.

True length Gg

$$= \sqrt{(XY + MR - pr)^2 + hg^2}$$
$$= \sqrt{(580 + 500 \cdot 5 - 303 \cdot 53)^2 + 1175 \cdot 3^2}$$
$$= 1408 \cdot 7 \text{ mm.}$$

Length of chords AB, BC, etc. = MR × ·522
$$= 500 \cdot 5 \times \cdot 522$$
$$= 216 \cdot 3 \text{ mm.}$$

Length of chords ab, bc, etc. = mr × ·522
$$= 350 \cdot 5 \times \cdot 522$$
$$= 183 \cdot 0 \text{ mm.}$$

PATTERN FOR TWO-WAY BRANCH

First mark off the centre line Aa (909·4 mm) of the pattern in a convenient position, and construct triangles each side of this line as shown in the diagram. Take true length Ba (977·4 mm), and strike arcs from centre a each side of Aa. Cut the arcs in B with distance AB (216·3 mm). Next describe arcs from B with true length Bb (944·6 mm), and cut in b from centre a with distance ab (183 mm). Take true length Cb (1066·5 mm) and draw arcs from b. Cut in C with distance BC (216·3 mm) from B.

The remainder of the pattern is drafted in the same manner.

POSITION OF JOINT LINE POINTS

Obtain the joint cut true lengths on the pattern as follows :—

True length Ee' $= \dfrac{\cdot 5MR}{\cdot 5MR + XY - \cdot 5pr} \times E e$

$$= \dfrac{250 \cdot 5}{250 \cdot 5 + 580 - 151 \cdot 76} \times 1288$$

$$= 475 \cdot 5 \text{ mm.}$$

True length Ff' $= \dfrac{\cdot 866MR}{\cdot 866MR + XY - \cdot 866pr} \times F f$

$$= \dfrac{433 \cdot 43}{433 \cdot 43 + 580 - 262 \cdot 86} \times 1376 \cdot 7$$

$$= 794 \cdot 6 \text{ mm.}$$

$$\text{True length } Gg' = \frac{MR}{MR + XY - pr} \times Gg$$

$$= \frac{500\cdot5}{500\cdot5 + 580 - 303\cdot53} \times 1408\cdot7$$

$$= 907\cdot6 \text{ mm.}$$

Mark these true lengths off on the pattern lines as shown. Take distance Gg' (907·6 mm) and measure from G on Gg, marking point g'. Next take Ff' (794·6 mm) and mark off point f' from F on Ff. Finally, mark point e' on Ee from E with true length Ee' (475·5 mm). Join all the points as shown to complete the branch pattern.

THREE-WAY JUNCTION WITH INCLINED INLETS

In the next example, which is of a three-way junction piece with inclined round inlets, as depicted in Fig. 62, methods of calculation give very satisfactory results for what is a rather complicated job to develop by the usual geometrical methods. The circular base is 350 mm inside diameter, to which three top inlets, each 200 mm inside diameter, are inclined at an angle of 30°. Vertical height to the top centres is 300 mm. The off-centre distance XY is 180 mm, and material used is 2 mm thick.

Calculate vertical heights ha, hb, etc. :—

$$\text{Vertical height } ha = hd - (\sin 30° \times mr)$$
$$= 300 - \cdot5 \times 101$$
$$= 249\cdot5 \text{ mm.}$$

$$\text{Vertical height } hb = hd - (\sin 30° \times \cdot866mr)$$
$$= 300 - \cdot5 \times 87\cdot47$$
$$= 256\cdot3 \text{ mm.}$$

$$\text{Vertical height } hc = hd - (\sin 30° \times \cdot5mr)$$
$$= 300 - \cdot5 \times 50\cdot5$$
$$= 274\cdot8 \text{ mm.}$$

$$\text{Vertical height } he = hd + (\sin 30° \times \cdot5mr)$$
$$= 300 + (\cdot5 \times 50\cdot5)$$
$$= 325\cdot3 \text{ mm.}$$

Vertical height $hf = hd + (\sin 30° \times ·866mr)$
$$= 300 + ·5 \times 87·47$$
$$= 343·7 \text{ mm.}$$

Vertical height $hg = hd + (\sin 30° \times mr)$
$$= 300 + ·5 \times 101$$
$$= 350·5 \text{ mm.}$$

Now obtain plan radius (pr).
$$pr = mr \times \cos 30°$$
$$= 101 \times ·866$$
$$= 87·47 \text{ mm.}$$

Find the pattern true lengths as follows :—

True length Aa
$$= \sqrt{(XY + pr - MR)^2 + ha^2}$$
$$= \sqrt{(175 + 87·47 - 176)^2 + 249·5^2}$$
$$= 264·0 \text{ mm.}$$

True length Ba
$$= \sqrt{(XY + pr - ·866MR)^2 + (·5MR)^2 + ha^2}$$
$$= \sqrt{(175 + 87·47 - 152·42)^2 + 88^2 + 249·5^2}$$
$$= 286·5 \text{ mm.}$$

True length Bb
$$= \sqrt{(XY + ·866pr - ·866MR)^2 + (·5MR - ·5mr)^2 + hb^2}$$
$$= \sqrt{(175 + 75·75 - 152·42)^2 + (88 - 50·5)^2 + 256·3^2}$$
$$= 277·1 \text{ mm.}$$

True length Cb
$$= \sqrt{(XY + ·866pr - ·5MR)^2 + (·866MR - ·5mr)^2 + hb^2}$$
$$= \sqrt{(175 + 75·75 - 88)^2 + (152·42 - 50·5)^2 + 256·3^2}$$
$$= 320·2 \text{ mm.}$$

True length Cc
$$= \sqrt{(XY + ·5pr - ·5MR)^2 + (·866MR - ·866mr)^2 + hc^2}$$
$$= \sqrt{(175 + 43·73 - 88)^2 + (152·42 - 87·47)^2 + 274·8^2}$$
$$= 311·2 \text{ mm.}$$

True length Dc

$$= \sqrt{(XY + \cdot 5pr)^2 + (MR - \cdot 866mr)^2 + hc^2}$$
$$= \sqrt{(175 + 43 \cdot 73)^2 + (176 - 87 \cdot 47)^2 + 274 \cdot 8^2}$$
$$= 362 \cdot 2 \text{ mm.}$$

True length Dd

$$= \sqrt{XY^2 + (MR - mr)^2 + hd^2}$$
$$= \sqrt{175^2 + (176 - 101)^2 + 300^2}$$
$$= 355 \cdot 3 \text{ mm.}$$

True length Ed

$$= \sqrt{(XY + \cdot 5MR)^2 + (\cdot 866MR - mr)^2 + hd^2}$$
$$= \sqrt{(175 + 88)^2 + (152 \cdot 42 - 101)^2 + 300^2}$$
$$= 402 \cdot 3 \text{ mm.}$$

True length Ee

$$= \sqrt{(XY + \cdot 5MR - \cdot 5pr)^2 + (\cdot 866MR - \cdot 866mr)^2 + he^2}$$
$$= \sqrt{(175 + 88 - 43 \cdot 73)^2 + (152 \cdot 42 - 75 \cdot 75)^2 + 325 \cdot 3^2}$$
$$= 399 \cdot 8 \text{ mm.}$$

True length Fe

$$= \sqrt{(XY + \cdot 866MR - \cdot 5pr)^2 + (\cdot 5MR - \cdot 866mr)^2 + he^2}$$
$$= \sqrt{(175 + 152 \cdot 42 - 43 \cdot 73)^2 + (88 - 87 \cdot 47)^2 + 325 \cdot 3^2}$$
$$= 431 \cdot 6 \text{ mm.}$$

True length Ff

$$= \sqrt{(XY + \cdot 866MR - \cdot 866pr)^2 + (\cdot 5MR - \cdot 5mr)^2 + hf^2}$$
$$= \sqrt{(175 + 152 \cdot 42 - 75 \cdot 75)^2 + (88 - 50 \cdot 5)^2 + 343 \cdot 7^2}$$
$$= 427 \cdot 6 \text{ mm.}$$

True length Gf

$$= \sqrt{(XY + MR - \cdot 866pr)^2 + (\cdot 5mr)^2 + hf^2}$$
$$= \sqrt{(175 + 176 - 75 \cdot 75)^2 + 50 \cdot 5^2 + 343 \cdot 7^2}$$
$$= 443 \cdot 2 \text{ mm.}$$

True length Gg

$$= \sqrt{(XY + MR - pr)^2 + hg^2}$$
$$= \sqrt{(175 + 176 - 87 \cdot 47)^2 + 350 \cdot 5^2}$$
$$= 438 \cdot 5 \text{ mm.}$$

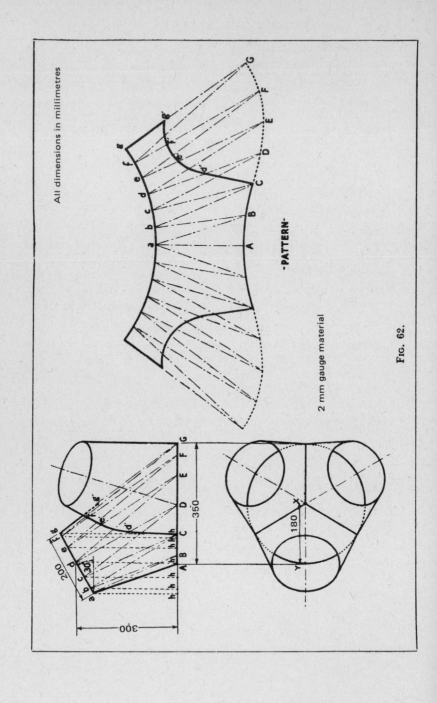

All dimensions in millimetres

-PATTERN-

2 mm gauge material

FIG. 62.

Length of chords AB, BC, etc. $= \text{MR} \times \cdot 522$

$$= 176 \times \cdot 522$$

$$= 91\cdot 8 \text{ mm.}$$

Length of chords ab, bc, etc. $= mr \times \cdot 522$

$$= 101 \times \cdot 522$$

$$= 52\cdot 7 \text{ mm.}$$

DRAFTING OF THREE-WAY PIECE PATTERN

Measure off in a suitable position the pattern middle line Aa (264 mm). Take true length Ba (286·5 mm), and describe arcs each side of Aa from centre a. With distance AB (91·8 mm) cut the arcs in B, using A as centre. Next take true length Bb (277·1 mm) and strike arcs from B. From centre a cut these arcs in b with distance ab (52·7 mm). Take true length Cb (320·2 mm), describe arcs from b and cut in C from B with distance BC. Continue the development as shown in the diagram until all the pattern triangles are complete.

OBTAINING JOINT CUT POINTS

Calculate the position on the pattern lay-out of the joint line between the branches by using the following formulæ :—

True length Dd' $= \dfrac{\cdot 5\text{MR}}{\cdot 5\text{MR} + \cdot 866(\text{XY} - \cdot 5774 \times mr)} \times \text{D}d$

$$= \dfrac{88}{88 + \cdot 866(175 - 53\cdot 32)} \times 355\cdot 3$$

$$= 165\cdot 3 \text{ mm.}$$

True length Ee' $= \dfrac{\cdot 866\text{MR}}{\cdot 866\text{MR} + \cdot 866[\text{XY} - (\cdot 5pr + \cdot 5mr)]} \times \text{E}e$

$$= \dfrac{152\cdot 42}{152\cdot 42 + \cdot 866[175 - (43\cdot 73 + 50\cdot 5)]} \times 399\cdot 8$$

$$= 274\cdot 0 \text{ mm.}$$

True length Ff' $= \dfrac{\text{MR}}{\text{MR} + \cdot 866[\text{XY} - (\cdot 866pr + \cdot 2887mr)]} \times \text{F}f$

$$= \dfrac{176}{176 + \cdot 866[175 - (75\cdot 57 + 29\cdot 12)]} \times 427\cdot 6$$

$$= 317\cdot 7 \text{ mm.}$$

$$\text{True length } Gg' = \frac{MR}{MR + XY - pr} \times Gg$$

$$= \frac{176}{176 + 175 - 87 \cdot 47} \times 438 \cdot 5$$

$$= 292 \cdot 9 \text{ mm.}$$

Measure the true lengths off on the appropriate pattern lines. First take Gg' (292·9 mm) and mark off point g' on Gg from G. Next measure from F on Ff true length Ff' (317·7 mm) marking f'. From E measure Ee' (274 mm) on Ee, marking e'. Finally, on Dd mark off d' from D with distance Dd' (165·3 mm). Join all the points to complete the branch pattern.

SPECIAL JUNCTION PIECES

Calculation methods can also be used for obtaining the patterns for junction pieces with more than three branches, by adaptations of the formulæ already given for obtaining joint cut points. As, however, such types of duct work are the exception rather than the rule, the chosen examples should prove of more value for practical use. Many special types of junction pieces, some examples of which are given in Chapter Six, can have their patterns successfully drafted by the use of suitable formulæ, although some pattern drafters may perhaps prefer to use geometrical methods for a speciality job.

Chapter Eighteen

PIPE AND TRANSFORMER CALCULATIONS

CALCULATION methods can be used for the development of patterns for articles such as transformers and connecting pieces which have one or more of their end sections elliptical in shape. Such a job is illustrated in Fig. 63, in which a round pipe is cut obliquely to fit the elliptical end section of a transformer, the base of which is square.

As it is a rather tedious process to calculate the spacings on the elliptical joint line between the parts, the easiest procedure is first to calculate and lay-out the pattern for the cut cylinder. In this way the joint spacings *ab*, *bc*, *cd*, etc., can be picked up with the dividers and transferred, as required, to the drafting of the transformer pattern.

The job has a base 1200 mm square, inside sizes, and its height at the vertical centre line is 750 mm. To this line is inclined the top centre line at an angle of 120°. Length of the top centre line is 625 mm and the top inside diameter is 700 mm. Material used is 2 mm thick.

The top portion of the job is drafted by the parallel-line method, the development lines for which are calculated in the following manner.

First find the angle of the joint cut across the pipe to the pipe base *a'g'*.

$$\text{Angle of pipe cut} = 90° - \frac{120°}{2}$$

$$= 30°.$$

Next find the lengths of the parallel lines *aa'*, *bb'*, and *cc'* by multiplying the base mean radius, ·866 mean radius, and ·5 mean radius, in each case by tan 30°, and *subtracting* each result from *dd'*.

$$\text{Length } aa' = dd' - (mr \times \tan 30°)$$

$$= 625 - (351 \times ·5774)$$

$$= 422·3 \text{ mm.}$$

$$\text{Length } bb' = dd' - (\cdot 866mr \times \tan 30°)$$
$$= 625 - (303 \cdot 97 \times \cdot 5774)$$
$$= 449 \cdot 4 \text{ mm}.$$
$$\text{Length } cc' = dd' - (\cdot 5mr \times \tan 30°)$$
$$= 625 - (175 \cdot 5 \times \cdot 5774)$$
$$= 523 \cdot 7 \text{ mm}.$$

Lengths of ee', ff', and gg' are obtained by multiplying $\cdot 5$ mean radius, $\cdot 866$ mean radius, and mean radius by $\tan 30°$, and *adding* each result to dd'.

$$\text{Length } ee' = dd' + (\cdot 5mr \times \tan 30°)$$
$$= 625 + (175 \cdot 5 \times \cdot 5774)$$
$$= 726 \cdot 3 \text{ mm}.$$
$$\text{Length } ff' = dd' + (\cdot 866mr \times \tan 30°)$$
$$= 625 + (303 \cdot 97 \times \cdot 5774)$$
$$= 800 \cdot 5 \text{ mm}.$$
$$\text{Length } gg' = dd' + (mr \times \tan 30°)$$
$$= 625 + (351 \times \cdot 5774)$$
$$= 827 \cdot 7 \text{ mm}.$$

As it is only necessary to develop a half-pattern of the pipe, the next procedure is to obtain the true length of half the base circumference. The most simple method of doing this is to multiply the mean radius by $3 \cdot 1416$.

$$\text{Half base circumference} = \text{mean radius} \times 3 \cdot 1416$$
$$= 3 \cdot 51 \times 3 \cdot 1416$$
$$= 110 \cdot 3 \text{ mm}.$$

METHODS OF DRAFTING PIPE PATTERN

Draw a horizontal line $110 \cdot 3$ mm in length in a convenient position. Divide it into six equal parts, and mark the points a', b', c', d', etc., as shown. Draw perpendiculars from the points and cut these off in lengths equal to aa', bb', etc. Thus the end line gg' is made 827 mm long, ff' $800 \cdot 5$ mm long, and so on. Letter the points found as shown, and join with an even curve to complete the half-pattern. The full pattern can be drafted by a repetition of the lines drawn.

CALCULATION OF TRANSFORMER TRUE LENGTHS

Before the pattern true lengths can be ascertained, the lengths of vertical heights ha, hb, hc, etc., must be calculated. The arithmetic involved is similar to that required for the pipe development. Obtain the lengths as follows :—

$$
\begin{aligned}
\text{Vertical height } ha &= hd - (mr \times \tan 30°) \\
&= 750 - (351 \times \cdot5774) \\
&= 547 \cdot 3 \text{ mm.}
\end{aligned}
$$

$$
\begin{aligned}
\text{Vertical height } hb &= hd - (\cdot866mr \times \tan 30°) \\
&= 750 - (303 \cdot 97 \times \cdot5774) \\
&= 573 \cdot 4 \text{ mm.}
\end{aligned}
$$

$$
\begin{aligned}
\text{Vertical height } hc &= hd - (\cdot5mr \times \tan 30°) \\
&= 750 - (175 \cdot 5 \times \cdot5774) \\
&= 648 \cdot 7 \text{ mm.}
\end{aligned}
$$

$$
\begin{aligned}
\text{Vertical height } he &= hd + (\cdot5mr \times \tan 30°) \\
&= 750 + (175 \cdot 5 \times \cdot5774) \\
&= 851 \cdot 3 \text{ mm.}
\end{aligned}
$$

$$
\begin{aligned}
\text{Vertical height } hf &= hd + (\cdot866mr \times \tan 30°) \\
&= 750 + (303 \cdot 97 \times \cdot5774) \\
&= 926 \cdot 6 \text{ mm.}
\end{aligned}
$$

$$
\begin{aligned}
\text{Vertical height } hg &= hd + (mr \times \tan 30°) \\
&= 750 + (351 \times \cdot5774) \\
&= 952 \cdot 7 \text{ mm.}
\end{aligned}
$$

The vertical heights can now be incorporated in the true length formulæ, which are similar to those given for the rectangular to round transformer described in Chapter Twelve.

$$
\begin{aligned}
\text{True length } Ba &= \sqrt{(HL - mr)^2 + ha^2} \\
&= \sqrt{(600 - 351)^2 + 547 \cdot 3^2} \\
&= 601 \cdot 3 \text{ mm.}
\end{aligned}
$$

$$
\begin{aligned}
\text{True length } Hb &= \sqrt{(HL - \cdot866mr)^2 + (HB - \cdot5mr)^2 + hb^2} \\
&= \sqrt{(600 - 303 \cdot 97)^2 + (600 - 175 \cdot 5)^2 + 573 \cdot 4^2} \\
&= 772 \cdot 4 \text{ mm.}
\end{aligned}
$$

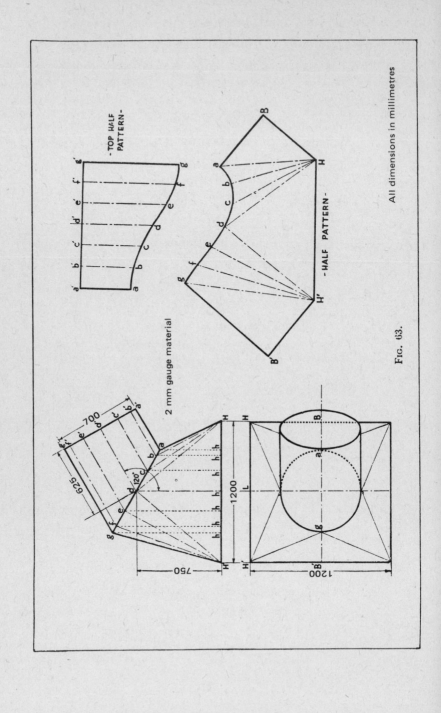

- TOP HALF PATTERN -

- HALF PATTERN -

2 mm gauge material

700

625

750

1200

1200

120°

All dimensions in millimetres

Fig. 63.

True length $Hc = \sqrt{(HL - \cdot 5mr)^2 + (HB - \cdot 866mr)^2 + hc^2}$

$\qquad = \sqrt{(600 - 175 \cdot 5)^2 + (600 - 303 \cdot 97)^2 + 648 \cdot 7^2}$

$\qquad = 829 \cdot 8$ mm.

True length $Hd = \sqrt{HL^2 + (HB - mr)^2 + hd^2}$

$\qquad = \sqrt{600 + (600 - 351)^2 + 750^2}$

$\qquad = 992 \cdot 2$ mm.

True length $H'd = $ True length Hd

$\qquad = 992 \cdot 2$ mm.

True length $H'e = \sqrt{(H'L - \cdot 5mr)^2 + (H'B' - \cdot 866mr)^2 + he^2}$

$\qquad = \sqrt{(600 - 175 \cdot 5)^2 + (600 - 303 \cdot 97)^2 + 851 \cdot 3^2}$

$\qquad = 996 \cdot 3$ mm.

True length $H'f = \sqrt{(H'L - \cdot 866mr)^2 + (H'B' - \cdot 5mr)^2 + hf^2}$

$\qquad = \sqrt{(600 - 303 \cdot 97)^2 + (600 - 175 \cdot 5)^2 + 926 \cdot 6^2}$

$\qquad = 1061 \cdot 3$ mm.

True length $H'g = \sqrt{(H'L - mr)^2 + H'B'^2 + hg^2}$

$\qquad = \sqrt{(600 - 351)^2 + 600^2 + 952 \cdot 7^2}$

$\qquad = 1153 \cdot 1$ mm.

True length $B'g = \sqrt{(H'L - mr)^2 + hg^2}$

$\qquad = \sqrt{(600 - 351)^2 + 952 \cdot 7^2}$

$\qquad = 984 \cdot 7$ mm.

DRAFTING THE TRANSFORMER PATTERN

Draw in a suitable position the seam line Ba (601·3 mm) and letter the points. From B draw a line at right angles to Ba and cut it off in length distance HB (600 mm). Join H to a to complete the first triangle. Take true length Hb (772·4 mm) and describe an arc from centre H. Cut this arc in b from centre a with spacing ab from the curve on the pipe lay-out. Next take true length Hc (829·8 mm) and strike an arc from centre H. With spacing bc from the pipe lay-out cut the arc previously drawn in c. Take true length Hd (992·2 mm), describe an arc from H, and cut in d with joint spacing cd. From centre d strike an arc with true length $H'd$ (992·2 mm), and cut in H' with base distance HH' (1200 mm) from centre H. The remainder of the half-pattern is drafted in a similar manner, and when complete the girth lines of the pipe and transformer half-patterns should be equal in length. Completion of the full pattern can be obtained by a repetition in reverse order of the lines already drawn.

THE FUTURE OF CALCULATION METHODS

The selected examples given in preceding chapters will furnish some indication of the possibilities of accurate pattern drafting by the use of mathematical formulæ. Ventilation work in particular offers many opportunities for standardisation of fitting design, thus allowing calculation methods to be used to the best advantage. It should present little difficulty to any skilled pattern drafter to evolve additional formulæ from those given in this book for many standard types of articles. No hard-and-fast rule can be laid down as to which method of development, geometrical or calculation, should be used in any given set of circumstances, as this obviously depends to a great degree upon the desired accuracy of the finished product and the ability of the pattern drafter. In many cases a combination of the two methods can be used successfully for development problems.

Chapter Nineteen

DETAIL FITTING CALCULATIONS

I N the large scale production of aircraft the fabrication of detail fittings has become an important part of the sheet metal industry. The fittings, of which a few examples have already been described in Chapter Eleven, are generally made from mild steel, aluminium, or duralumin. Accuracy in manufacture is essential, as aircraft drawings specify limits which are usually only obtainable by using accurately developed templates in conjunction with adequate tooling. For quantity output, uniform quality of the finished product, whether of welded or riveted construction, is ensured by the employment of suitable jigs and fixtures.

One of the most important features of pattern development for precision sheet metal work is the correct calculation of bend allowances. For such laying-out work a combination of both geometrical and calculation methods can give very satisfactory results, particularly in the case of jobs for which the patterns have to be triangulated.

THE USE OF BEND ALLOWANCES

Before the development of the surface of an article can be obtained, it is necessary to calculate the exact length of the edges of the metal as they would appear when set-out in the flat. The net pattern is then drafted and, if flanges have to be added, suitable allowances are made to the pattern profile. As an aid to the calculation of bend allowances it is a sound idea to sketch first an enlarged end or plan view of the fitting—not necessarily to scale—showing the metal thickness, and dimension it from the working drawing.

In Fig. 64 is shown the development of an attachment bracket made from material 2 mm thick.

The developed length of the fitting may be calculated in the following manner : —

(Note that as no inside bend radius is given, this will be reckoned as 2G, *i.e.*, twice the material thickness.)

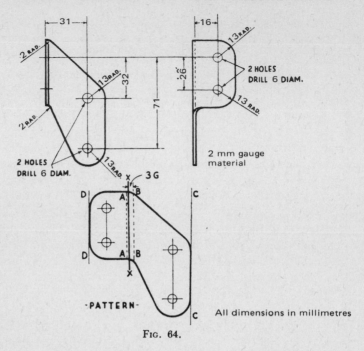

2 HOLES
DRILL 6 DIAM.

2 mm gauge
material

-PATTERN-

All dimensions in millimetres

FIG. 64.

First draw a similar sketch to Fig. 65, letter the vertical flat AD, as shown, and mark dimension 29 mm (16 + 13 mm radius). This measurement is from edge point D to the *inside* of the metal thickness. Next, mark 44 mm (31 + 13 mm radius) along the horizontal line from X, and mark flat BC.

Now calculate the mean line of the metal thickness for the stretched-out length of the bracket.

First find the length of the flat portion AD by deducting 2G (4 mm) from 29 mm, which gives 25 mm. Next subtract 3G (6 mm) from 44 mm to obtain 38 mm as the length of flat BC. Finally, obtain the bend allowance B.A. by first multiplying the mean radius $(2G + \frac{1}{2}G)$ by 90 (the number of degrees in the bend angle), and then multiplying by ·01745, as follows :—

$$
\begin{aligned}
\text{B.A.} &= \text{Mean radius} \times 90 \times \cdot01745 \\
&= 2 \cdot 5G \times 90 \times \cdot01745 \\
&= 5 \times 90 \times \cdot01745 \\
&= 7 \cdot 86 \text{ mm.}
\end{aligned}
$$

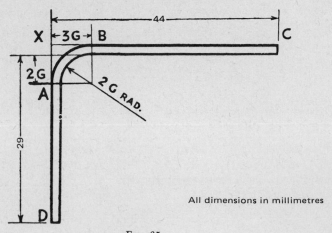

FIG. 65.

To start the pattern, first draw the elevation profile as shown in Fig. 64. Measure back from the profile line X-X a distance of 3G (6 mm), and draw a datum line B-B. Alternatively, measure flat BC (38 mm) from edge line C-C.

From B-B measure off the bend allowance (7·86 mm), marking the line A-A. At right angles to X-X draw two lines of indefinite length, and measure off from A the length of the flange AD (25 mm). To complete the pattern, mark off the radius corners and hole centres as shown. The bend line, which, for clarity, is not shown on the pattern, is drawn in the centre of the bend allowance. A simpler method of setting-out bend allowances, which is more adaptable for some types of fittings, will be described later.

CONTROL CABLE COVER

A typical example of an aircraft cover or guard is shown in Fig. 66. It is made from material 2 mm thick, and its purpose is to protect cables from damage. The pattern is marked-out by the triangulation method; but before the development is tackled it is essential to obtain the correct lengths in the flat of the edges of the job, as shown in the end view. To obtain these measurements, first sketch the rear and front edges, as shown at (a) and (b) respectively in Fig. 67. At this stage do not take

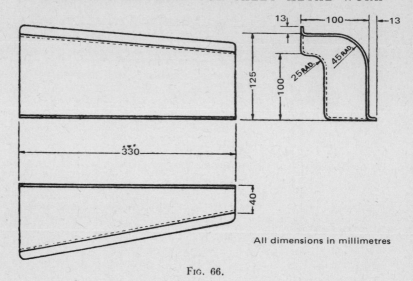

All dimensions in millimetres

FIG. 66.

into consideration the flanges on the job. These are dealt with separately, as they are added to the pattern after it has been fully triangulated. This method makes for simplification of the drafting process, and ensures that the correct bending allowance is added parallel to the appropriate pattern profile line.

OBTAINING THE BEND ALLOWANCES

First deal with the rear edge, as shown in Fig. 67 (a). The horizontal or minor leg is marked 40 mm long from the plan (Fig. 66). The length of flat 00′ is obtained by adding the inside bend radius to the metal thickness, and deducting from the minor leg dimension as follows :—

$$\text{Flat } 00' = 40 - (25 + 2)$$
$$= 13 \text{ mm}.$$

The flat portion 34 of the vertical, or major leg, is found in a similar manner.

$$\text{Flat } 34 = 100 - (25 + 2)$$
$$= 73 \text{ mm}.$$

The lengths of spacings 0-1, 1-2, and 2-3 can be found by dividing the quadrant 0-3 into three equal parts. These spacings are used in conjunction with true length lines to build up the pattern triangles.

$$\text{Spacings 0-1, etc.} = \frac{\text{Mean radius} \times 90 \times \cdot 01745}{3}$$

$$= \frac{(25+1) \times 90 \times \cdot 01745}{3}$$

$$= 13 \cdot 6 \text{ mm.}$$

A more efficient method of obtaining the length of quadrant spacings is to multiply the quadrant mean radius by ·522. This gives a spacing of 13·57 mm and for all practical purposes 13·6 mm will ensure the correct length of the arc between the pattern triangle base points.

Next obtain the lengths of the front edge, as in Fig. 67 (b).

$$\text{Flat AA}' = 100 - (45+2)$$
$$= 53 \text{ mm.}$$
$$\text{Flat DE} = 125 - (45+2)$$
$$= 78 \text{ mm.}$$
$$\text{Spacings A-B, etc.} = \text{Mean radius} \times \cdot 522$$
$$= (45+1) \times \cdot 522$$
$$= 24 \text{ mm.}$$

The stretched-out lengths of the flange, shown enlarged at Fig. 67 (c), are found thus :—

$$\text{Flat } ad = 13 - 2G$$
$$= 13 - 4$$
$$= 9 \text{ mm.}$$

Bend allowance :—

$$\text{B.A.} = \text{Mean radius} \times \text{bend angle} \times \cdot 01745$$
$$= 2 \cdot 5G \times 90 \times \cdot 01745$$
$$= 5 \times 90 \times \cdot 01745$$
$$= 7 \cdot 86 \text{ mm.}$$

Fig. 67 (d) shows the method of marking off the flange allowances to the pattern profile line E-4. Further reference will be made to this later.

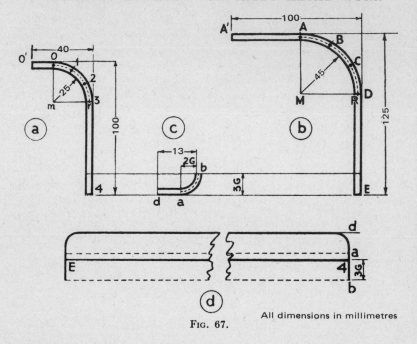

FIG. 67.

All dimensions in millimetres

GEOMETRICAL CONSTRUCTION FOR THE PATTERN

Geometrical methods are now to be used to obtain the pattern development. This is drafted by the lay-out system of triangulation, as depicted in Fig. 68. First draw an elevation in a suitable position and mark off on the vertical edge lines A'-E' and 0'-4' the shape of the end surfaces of the job. Note that the perimeters A'-E and 0'-4 denote the centre line of the metal thickness, and must be measured off accordingly. For instance, E-E' is 99 mm long, obtained by deducting half the metal thickness from 100 mm.

After marking off the elevation and end profiles, divide the quadrants A-D and 0-3 into three equal parts and draw horizontal lines (shown dotted) to the elevation edge lines from the points found. These lines are ordinates required for the lay-out construction. Next join opposite edge-points, as shown by chain-dotted lines, to form triangles. Set-out a layout in a convenient position by first drawing a vertical line of indefinite length and marking a point 0'. From 0' measure off 00' and ordinates 1, 2, and 3 from the elevation, and mark as shown. Next, from the same point, mark off AA' and ordinates B, C, and D, draw horizontal lines of indefinite length from the points, and mark each with the appropriate letter.

TRIANGULATING THE PATTERN

First take edge line 0'-A' from the elevation, and draw it in a convenient position. Now measure the same line along the top line of the lay-out from point 0' and mark A'. Connect the point to 0 on the lay-out perpendicular, and with this true length describe an arc from point A' on the pattern line. Cut this arc in 0 from centre 0' with distance 00' (13 mm). Next, with the same elevation line 0'-A', measure along lay-out line A, and from the point found join to 0 on the perpendicular. With this true length describe an arc from 0 in the pattern, and cut in A with distance AA' (53 mm) from centre point A'.

Now take chain-dotted elevation line 0'-B and measure it along lay-out line B. Join the point marked to 0 on the perpendicular. Take this true length and strike an arc from 0 in the pattern. With distance A-B (24 mm), using A as centre, cut the arc just drawn in B. Pick up 1-B from the elevation, measure it along lay-out line B and join to perpendicular point 1. Describe an arc with the true length from centre B in the pattern to meet in 1 an arc drawn from the point 0 with distance 0-1 (13·6 mm). Continue the development on similar lines, marking off in turn each elevation line on its appropriate lay-out line, and triangulating to the correctly numbered point on the perpendicular. The true lengths thus found are then used to build up the remaining pattern triangles as shown. Finally, join up all points with smooth curves.

ADDING FLANGES TO THE PATTERN

When the net pattern is complete the flanges are added in the following manner: From the profile lines 0'-A' and E-4 measure back a distance of 3G (6 mm), and draw parallel datum lines to the profile lines, as shown at (d) in Fig. 67. Measure from each datum line the bend allowance B.A. (7·9 mm), and then add the length of each flat portion ad (9 mm) to complete the pattern.

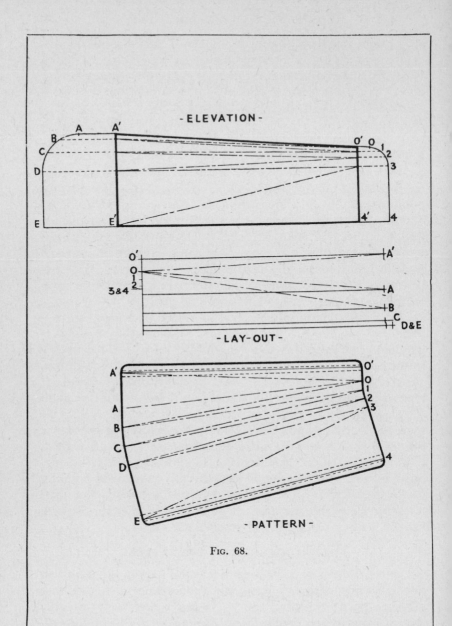

FIG. 68.

Chapter Twenty

FAIRING DEVELOPMENTS

PRODUCTION methods have been revolutionised in the sheet metal industry by the extensive use of machinery for deep drawing and pressing. The plastic deformation of sheet material to almost any desired shape by such means has only to a limited extent made obsolete the working-up of double curvature forms by hand methods. Uniform quality articles can be drawn or pressed for quantity output at comparatively low cost, yet despite these advantages there will always be ample scope for the skilled sheet metal worker to use hand processes, particularly in the case of " short runs " where the cost and time necessary to tool up for pressing or deep drawing cannot be justified. Modifications to design also often call for methods that will keep production flowing until such time as pressings can be obtained.

As an example of work which could either be pressed or hand fabricated, the fairing shown in Fig. 69 is typical of many met with in the aircraft industry. This job, which is made of material 1 mm thick, although complex in form can be made up quite easily in two portions with welded joints. The body portion is cut out and shaped up in one piece, and the flange at the top " thrown-off " in a suitable former or jig, after welding the seam. The double curvature portion is beaten up by hand with a pear-shaped mallet into a suitably shaped wooden former.

Fig. 69 depicts the plan, elevation, and end view of the fairing, the most important dimensions being shown. The pattern for the body portion of the job is triangulated in one piece, the seam for welding being made at the front. Before tackling the development of the pattern, allowances for the flange must be calculated, taking into consideration the thickness of the material.

Referring to Fig. 69 it will be seen that the flange round the top of the object, with the exception of the rear end, is

177

9 mm in width. Calculate the bend allowance and the length of the flat portion as follows :—

$$B.A. = MR \times \text{bend angle} \times \cdot 01745$$
$$= 2\cdot 5G \times 90 \times \cdot 01745$$
$$= 2\cdot 5 \times 90 \times \cdot 01745$$
$$= 3\cdot 9 \text{ mm.}$$

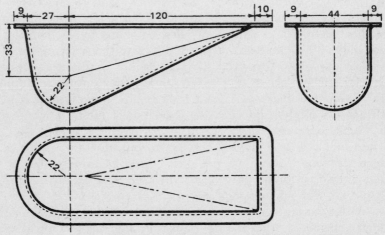

FIG. 69.   All dimensions in millimetres

Next deduct 3G (3 mm) from the width, 9 mm, to obtain 6 mm as the length of the flat portion.

Allowances for the rear flange are rather more difficult to obtain, as before the bend allowance can be found it is necessary to calculate the bend angle. The method of doing this is shown in Fig. 70 at (a) and (b). It is necessary first of all to calculate the angle—marked BAD—at which the bottom of the job is inclined to the flanged top. Working from the given dimensions as at Fig. 70 (a), first calculate the length of the hypotenuse AC as follows :—

$$\text{Length AC} = \sqrt{AB^2 + BC^2}$$
$$= \sqrt{120^2 + 33^2}$$
$$= 124\cdot 5 \text{ mm.}$$

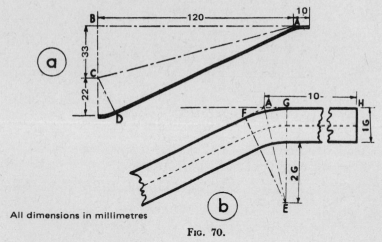

All dimensions in millimetres

Fig. 70.

Now find angle BAC :—

$$\cos \text{ angle } BAC = \frac{AB}{AC}$$

$$= \frac{120}{124 \cdot 5}$$

$$= \cdot 9638$$

hence,   angle $BAC = 15° \ 30'$

Next obtain angle CAD :—

$$\sin \text{ angle } CAD = \frac{CD}{AC}$$

$$= \frac{22}{124 \cdot 5}$$

$$= \cdot 1606$$

hence,   angle $CAD = 9° \ 15'$.

Thus angle BAD is equal to the sum of the two angles just obtained, *i.e.*, $15° \ 30' + 9° \ 15' = 24° \ 45'$.

Refer now to Fig. 70 (*b*), which shows an enlarged view of the flange.

Bend angle FEG is similar to angle BAD, proof of which is as follows :—

Angle EAF is found by deducting $24° \ 45'$ from $180°$, and dividing by two.

$$\text{Angle EAF} = \frac{180 - 24° \ 45'}{2}$$
$$= 77° \ 38'.$$

Angle AEF, which is similar to angle AEG, is obtained by deducting 77° 38' from 90° :—

$$\text{Angle AEF} = 90° \ -77° \ 38'$$
$$= 12° \ 22'.$$

Thus the bend angle FEG = 12° 22' × 2 = 24° 44'.
Next calculate the bend allowance :—

$$\text{B.A.} = 2 \cdot 5G \times 24° \ 44' \times \cdot 01745$$
$$= 2 \cdot 5 \times 24 \cdot 733 \times \cdot 01745$$
$$= 1 \cdot 08 \text{ mm}.$$

It is now necessary to obtain the length of flat GH, which is calculated by deducting AG from AH.

Find the length of AG, which is similar in length to AF, as follows :—

$$\text{AG} = \tan \text{AEG} \times \text{EG}$$
$$= \tan 12° \ 22' \times 3G$$
$$= \cdot 66 \text{ mm}.$$

Therefore the length of flat GH is equal to 10 mm − ·66 mm, which is 9·33 mm.

Next calculate the length of spacings round the semicircular end of the base of the job. In the development of the fairing shown in Fig. 71 the quadrant 0-3 on the edge line 0-A' is divided into three equal parts 0-1, 1-2, and 2-3 as shown.

$$\text{Spacings} = \text{mean radius} \times \cdot 522$$
$$= (22 + \cdot 5) \times \cdot 522$$
$$= 11 \cdot 74 \text{ mm}.$$

The method explained for calculating a bend angle other than 90° has been dealt with in detail, as the method used is applicable to many types of bend allowances for sheet metal aircraft work. It will be noticed that the dimensions involved are very small, and may give rise to the query that they are not worth while taking into consideration during the bending process. Experience proves, however, that time spent working out the correct allowances for portions of the job which must match up with other components is well repaid when the template is put into service.

CONSTRUCTION OF ELEVATION AND LAY-OUT

Having obtained all the necessary bending allowances, the drafting of the body pattern can next be carried out in the manner shown in Fig. 71.

First draw an elevation profile of the job to the mean line dimensions. On the top edge describe a quadrant using a top mean radius of 22·5 mm (22 + ·5 mm) as shown. Divide the quadrant into three equal spacings, mark the points 0, 1, 2, and 3, and draw ordinates to the top edge line. Next construct a similar quadrant on the joint line A-d, mark the points A, B, C, and D, and draw in the ordinates. Finally, repeat a similar construction on line 0'-d. Join all the points between edge lines and corner A'a to form the elevation triangles.

Now tackle the lay-out for obtaining the true pattern lengths of the elevation lines. Erect a vertical line of indefinite length in a suitable position, and mark a point 0. From this point measure off ordinates 1, 2, and 3, and mark the points. In this example the top and bottom ordinates are the same length, so draw horizontal lines of indefinite length from the points and mark them A, B, C, A' and D, as shown.

DRAFTING THE FAIRING BODY PATTERN

In a suitable position first mark off the elevation edge line 0'-A'a, and mark the points 0 and a. Next measure this line along the lay-out line A' and D and connect the point to 0 on the perpendicular. Using point 0 on the pattern line as centre, describe an arc with this true length, and cut it in A' with the half inside width of the job A'-a (22 mm) from centre a. Take line A'-1 from the elevation (shown chain-dotted), measure off on lay-out line A', and join to 1 on the lay-out perpendicular. With this true length describe an arc from centre A' in the pattern, and cut in 1 from centre 0 with spacing 0-1.

Now take A'-2, measure it along lay-out line A' and tri-angulate to 2 on the perpendicular. From point A' in the pattern describe an arc with the true length and cut it in

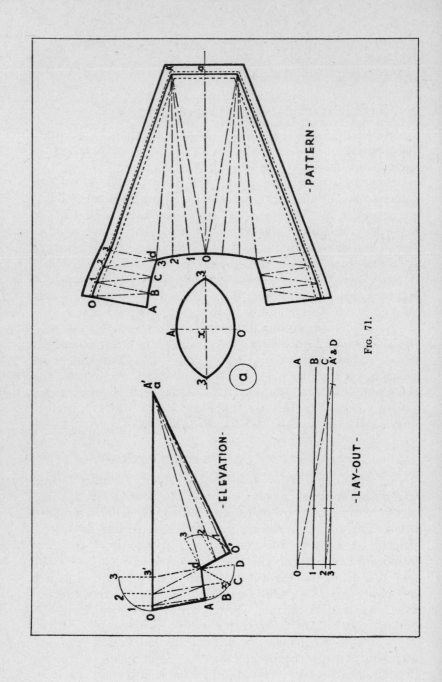

- PATTERN -

- ELEVATION -

- LAY-OUT -

FIG. 71.

2 from centre 1 with spacing 1-2. Now take elevation length A'-3 (the latter point is also marked d), strike an arc from pattern centre point A', and cut it in 3 from centre 2 with spacing 2-3.

To construct the flat triangular side of the job take 3'-d from the elevation, and describe an arc from centre 3d. With elevation distance A'-3' cut the arc previously drawn in 3 from centre A'.

For the triangulation of the front portion of the fairing take line D-2 from the elevation, step it along lay-out line A' and D and join to point 2 on the perpendicular. From point 3d in the pattern describe an arc with this true length, and cut it in 2 from centre 3 with spacing 2-3 (11·74 mm). Take line 2-C, measure it off on lay-out line C and extend to 2 on the perpendicular. Strike an arc from 2 in the pattern, and cut it in C with a spacing described from centre 3d. Continue the remainder of the half-pattern development in a similar manner.

Finally, draw in the half-pattern profile lines and measure back an equidistant line a distance of 3 gauge (3 mm) from edge line 0-A'. From this datum line measure off the bend allowance (1·08 mm) and then the flat portion (6 mm) in the manner as described for the previous example. Next, from profile line A'-a measure back a parallel datum line a distance of ·66 mm—AF in Fig. 70 (b)—then measure off the bend allowance (1·08 mm), and, lastly, the flat GH (9·33 mm).

The full pattern is drafted by reproducing in reverse order from line 0-a the lines already drawn. If desired, the whole pattern can be drafted and the flange allowances added.

To draft the pattern for the double curvature portion, first bisect arc A-0' in the elevation at x and join to d. Then draw an indefinite horizontal line in a convenient position and mark a point x as in Fig. 71 (a). The distances x-3 each side of the centre point are equal in length to three spacings, i.e., 35·22 mm (11·74 mm × 3). Draw a perpendicular through point x. Take chords A-x and x-0' from the elevation and mark off on the pattern perpendicular from centre x, as shown. With radius A-0 and centre 0 describe an arc through A to 33, and complete the pattern by describing a similar arc through point 0 from centre A.

FABRICATION OF THE FAIRING

After cutting out the patterns, the body is shaped up, the rounded front halves brought together, and the seam 0-A welded. The job must then be placed in a suitably shaped former and the flanges " thrown off " at the appropriate bending lines. A certain amount of stretching by malleting is necessary at the flange corners, as there is insufficient material at these points for their correct formation. The round portion of the flange at the front of the job must also be carefully stretched over until the whole flange lies level. It will be noted that the body is inclined at an angle to the flange at point 0, but the stretching process at this point compensates for any slight dimensional discrepancy using the bending allowance for a 90° bend.

Finally, the double curvature portion is beaten up to shape, trimmed to size where necessary, inserted in position in the fairing, and welded. Individual fitting is always necessary, because it is only possible to develop an approximate pattern for double curvature work, mainly on account of variation in the amount of stretching that the material undergoes during the beating-up process.

Any welding stresses set-up in the fitting can be relieved by judiciously hammering the joints on a suitably shaped planished steel head. Distortion in this particular type of job, in any case, will be only very slight, but on many structures a great deal of twisting can result from the use of oxy-acetylene welding.

Difficulties sometimes arise through dimensional distortion of the welded component when welding fixtures are designed directly from the information given on the working drawing, so from a practical point of view it is often a decided advantage before tooling a job for repetition work to make a " first-off," solely by hand, from the developed template. It is then possible to observe the degree of distortion set-up in the structure, and to allow for this in the design of the welding fixtures.

FABRICATION OF A RUDDER HEEL FAIRING

Innumerable types of aircraft fairings made on similar lines to the last example can be satisfactorily fabricated by hand methods. A good example of this class of work is given in Fig. 72, which shows a working drawing of a rudder heel fairing made from material 1 mm thick. The body of the job is partly conical in shape, with the front portion rounded off to a double curvature form. For ease of fabrication the body is shaped up in one piece, with a welded seam, as shown at 0-A in Fig. 74. A double curvature piece is welded into position on joint lines 0-d' and a-d'.

It is most essential, in order to obtain a correct fit to the rudder frame, that the base and end of the fairing are made dimensionally correct. Therefore the true lengths of the edges of the material must be calculated before the drafting of the body pattern can be tackled.

CALCULATIONS FOR METAL THICKNESS

Shown in Fig. 73 at (*a*) and (*b*) respectively are the methods for obtaining the true lengths of the base and end from the given dimensions on the working drawing. It is only necessary to work out the half-perimeters of the base and end, because each half of the job is symmetrical on the plan horizontal centre line. Draw a sketch of the base and end half-sections, and divide the surfaces of both into convenient right-angled triangles as shown. First tackle the calculations for the half-base as at (*a*) in the following manner.

Obtain the length of the hypotenuse EF as follows :—

$$\text{Length EF} = \sqrt{FG^2 + EG^2} \quad (FG = 536 \cdot 7 - 75 \cdot 2$$
$$= 461 \cdot 5 \text{ mm})$$

$$= \sqrt{461 \cdot 5^2 + 16 \cdot 3^2}$$
$$= 461 \cdot 8 \text{ mm.}$$

Next find angle EFG :—

$$\cos \text{ angle EFG} = \frac{FG}{EF}$$

$$= \frac{461 \cdot 5}{461 \cdot 8}$$

$$= \cdot 99913$$

hence, angle EFG $= 2° 13'$.

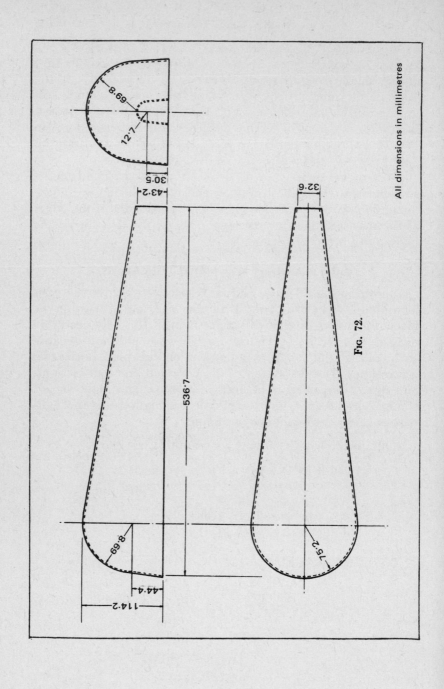

Fig. 72.

All dimensions in millimetres

Then obtain angle DFE :—

$$\cos \text{ angle DFE} = \frac{\text{DF}}{\text{EF}}$$

$$= \frac{75 \cdot 2}{461 \cdot 8}$$

$$= \cdot 1626$$

hence, angle DFE $= 80° \, 38'$.

Note that DF is equal to the bend inside radius (75·2 mm).
From 180° next deduct the sum of angles EFG and DFE
for the bend angle.

$$\text{Bend angle AFD} = 180° - (2° \, 13' + 80° \, 38')$$

$$= 97° \, 09'.$$

Next find the bend allowance :—

$$\text{B.A.} = \text{MR} \times 97° \, 09' \times \cdot 01745$$

$$= 75 \cdot 7 \times 97 \cdot 15 \times 01745$$

$$= 128 \cdot 33 \text{ mm.}$$

Length of spacings A-B, B-C, and C-D $= \dfrac{\text{B.A.}}{3}$

$$= \frac{128 \cdot 33}{3}$$

$$= 42 \cdot 77 \text{ mm.}$$

Finally, obtain the length of flat DE :—

$$\text{Flat DE} = \sin \text{DFE} \times \text{EF}$$

$$= \sin 80° \, 38' \times 461 \cdot 8$$

$$= 455 \cdot 64 \text{ mm.}$$

After working out the perimeter of the half-base edge,
calculate the length of the half-end perimeter as follows.
First obtain the length of the hypotenuse EJ as at Fig.
73 (b) :—

$$\text{Length EJ} = \sqrt{\text{HJ}^2 + \text{EH}^2}$$

$$= \sqrt{30 \cdot 5^2 + 16 \cdot 3^2}$$

$$= 34 \cdot 58 \text{ mm.}$$

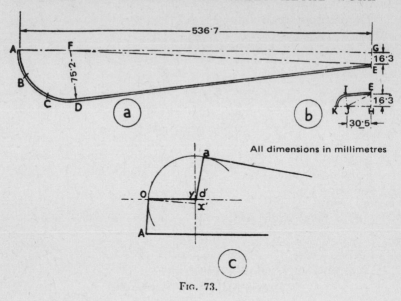

All dimensions in millimetres

Fig. 73.

Now find angle EJH :—

$$\cos \text{ angle } EJH = \frac{HJ}{EJ}$$

$$= \frac{30 \cdot 5}{34 \cdot 58}$$

$$= \cdot 882$$

hence,   angle EJH = 28° 07′.

Next obtain angle EJI :—

$$\cos \text{ angle } EJI = \frac{IJ}{EJ}$$

$$= \frac{12 \cdot 7}{34 \cdot 58}$$

$$= \cdot 3673$$

hence,   angle EJI = 68° 27′.

Note that IJ is equal to inside bend radius (12·7 mm).

The bend angle is found by subtracting the sum of angles EJH and EJI from 180° as follows :—

$$\text{Bend angle IJK} = 180° - (28° \ 07' + 68° \ 27')$$
$$= 83° \ 26'.$$

Now find the bend allowance :—

$$\text{B.A.} = \text{MR} \times 83° \ 26' \times \cdot01745$$
$$= 13\cdot2 \times 88\cdot43 \times \cdot01745$$
$$= 19\cdot22 \text{ mm.}$$

Finally, find the length of flat EI :—

$$\text{Flat EI} = \sin \text{EJI} \times \text{EJ}$$
$$= \sin 68° \ 27' \times 34\cdot58$$
$$= 32\cdot16 \text{ mm.}$$

SETTING-OUT FOR ELEVATION

The construction required for the front of the job when setting-out the elevation is depicted in Fig. 73 (c). Both top and front edge lines are tangential to an arc 0-x' radius described from centre x'. Lines are drawn to points 0 and a at right angles to the edge lines from centre point x'. From point 0 a line parallel to the base edge is drawn to cut line a-x' in d'. Thus the joint lines for the double curvature form are 0-d' and a-d' respectively. Line 0-d' intersects the vertical centre line at y.

Refer now to the elevation, which is drawn to mean line dimensions, in Fig. 74. From point d' a perpendicular is erected and cut in point 3 with an arc described from centre y with radius 0-y. The arc is divided into three equal spacings 0-1, 1-2, and 2-3. Perpendiculars (shown dotted) are drawn from the points to the joint line 0-d'. With radius a-x' from centre x' an arc is described, and a perpendicular drawn from d' to cut it in d. Arc a-d is divided into three equal spacings a-b, b-c, and c-d. Perpendiculars are drawn back to joint line a-d' from the points.

The half-base profile is constructed on edge line A-e. Arc A-D is divided into three equal parts A-B, B-C, and C-D, and perpendiculars drawn from the points to A-e. On the end

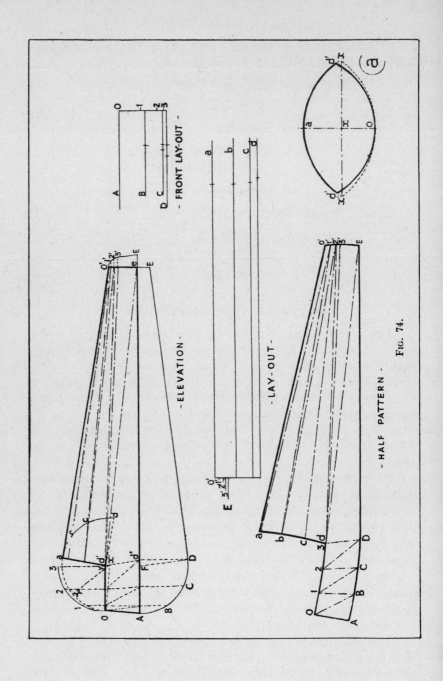

- FRONT LAY-OUT -

- ELEVATION -

- LAY-OUT -

- HALF PATTERN -

FIG. 74.

edge line 0'-*e* is constructed the half-end profile. Distance E-3' is made equal in length to flat EI (32·16 mm). Arc 0'-3', which is equal to arc IK in Fig. 73 (*b*), is divided into three equal spaces 0'-1', 1'-2', and 2'-3'. Perpendiculars to line 0'-*e* are drawn from the points. The elevation surface is divided into triangles with lines connecting all the edge points as shown.

LAY-OUT CONSTRUCTIONS

Two lay-outs are required for the drafting of the body pattern, the front lay-out being constructed as follows :—

First draw a vertical line and mark the point 0. From this point measure off ordinates 0, 1, 2, and 3 from the elevation, and also ordinates B, C, and D. Draw horizontal lines of indefinite length from the latter points, and mark each line A, B, C, and D as shown. To construct the lay-out for the body triangles, draw a vertical line, mark a point 0', and measure from this point ordinates 1', 2', and 3' from edge line 0'-*e*. The distance E-*e* is also measured from point 0', the point being marked E. Now, from point 0' measure off ordinates *b*, *c*, and *d*, draw indefinite horizontal lines from the points found and letter the lines *a*, *b*, *c*, and *d*.

DRAFTING THE HALF-PATTERN

To draft the half-pattern, first draw a line DE (455·64) in a convenient position. Pick up elevation line *e*-*d*' and measure it along lay-out line *d*. Join the point found to E on the perpendicular, and with this true length describe an arc from centre E on the pattern line. Take elevation length D-3, marked *d*'*d*" at the edge points, measure it along lay-out line D in the front lay-out, and join up to point 3 on the perpendicular. With this true length describe an arc from point D on the pattern line to cut the arc previously drawn from centre E in 3*d*.

Now take chain-dotted elevation line 2-D, measure it off on the front lay-out line D, and connect to point 2 on the perpendicular. Strike an arc from D in the half-pattern, and cut it in 2 from centre 3*d* with spacing 2-3 from the elevation. Pick up 2-C from the elevation, mark it off along front lay-out line C, and triangulate to 2 on the perpendicular. From point 2 in the pattern describe an arc and cut it in C with spacing C-D (42·74 mm) from centre D. Take elevation length 1-C, mark it off on lay-out line C, and join up to 1 on the perpendicular. To add this line to the half-pattern describe an arc from point C, and cut it in point 1 with spacing 1-2 from the elevation. Triangulate the remainder of the front of the half-pattern on the same lines.

Next take elevation line d'-$3'$, mark it along lay-out line d, and join to point $3'$ on the perpendicular. From point $3d$ in the half-pattern describe an arc with the true length. Now take distance E-$3'$, which is equal in length to EI (32·16 mm) in Fig. 73 (b), and from point E describe an arc, cutting the arc previously drawn in point $3'$. Take elevation length d'-2, mark it along lay-out line d, and triangulate to $2'$ on the perpendicular. Strike an arc from $3d$ in the half-pattern with the true length, and cut it in point $2'$ with distance $2'$-$3'$. This spacing is 6·41 mm in length, which is one-third of the bend allowance (19·22 mm) for angle IJK in Fig. 73 (b). Now pick up elevation length c-$2'$, step along lay-out c, and connect up to point $2'$ on the perpendicular. From centre $2'$ in the half-pattern strike an arc, and cut it in c from $3d$ with spacing c-d from the elevation. Draft the remainder of the half-pattern on similar lines. The whole pattern can then be obtained by reproducing, in reverse order, the lines already drawn.

DRAFTING THE DOUBLE CURVATURE PATTERN

The approximate pattern for the double curvature form is obtained by first bisecting arc 0-a in the elevation at x. Then join x to centre point x'. Next set-out in a suitable position, as at Fig. 74 (a), a horizontal line of indefinite length, and mark the centre point x. Draw a perpendicular through this point, and cut it off each side of x to the length of chords x-a and x-0 from the elevation. Make horizontal distances x-x', each side of x equal in length to elevation arc a-d. Join up the points with a smooth curve as shown. Finally, measure distance x'-d' from the elevation, and measure off on the pattern profile from points x'. The profile is adjusted to points d' as shown, but it must be clearly understood that as the pattern is only approximate it is necessary to trim off any surplus material after the shaping process has been completed.

FABRICATING THE FITTING

Cut out the patterns and shape up the body portion on a wooden former made to the internal dimensions of the fitting. Then place the job on a suitable welding fixture, and weld up the seam at 0-A. Shape up the pattern shown at (*a*) into a double curvature form by malleting in a suitable wooden former, and fit it into its correct place after trimming to size where necessary. Tack the whole job together on the welding fixture, and after welding planish the joints. The whole component can then be checked to the shape and sizes specified in the working drawing.

ALTERNATIVE BEND ALLOWANCE FORMULÆ

Before passing on to the next example it may be advisable to point out that throughout this work the centre line of the metal thickness has been taken as the neutral line. Experience has proved that this method is correct for all practical purposes, although cognisance must be taken of the practice in some factories of using bend allowance formulæ which assume that the neutral line is situated one-third to two-fifths of the metal thickness from the inside radius of the bend. Such formulæ allow for the fact that in bends of small radii the outside of the bend tends to stretch slightly more than the inside portion shrinks.

It is, however, impossible to be definite on this point, because so many different factors enter into the conditions when working the metal to shape. For instance, metal which has been bent by hammering on a bend block stretches to a larger extent than similar work flanged by press tools. Again, such considerations as the nature of the material used, its malleability, and other characteristics, are all variable factors which tend to cancel out any slight advantages which may be claimed for the use of empirical formulæ.

It will be observed in all the detail fitting development examples that the inside bend radii are always taken to be twice the thickness of material, but in those cases where the working drawing specifies definite bend radii dimensions these must be accounted for when calculating the lengths of mean lines.

Chapter Twenty-one

BEND LINE CALCULATIONS

A S previously explained, marking out templates for precision sheet metal work always involves consideration for the thickness of material used. Simple geometrical forms can easily be developed by the craftsman of average ability, but work of an accurate nature, such as aircraft detail fittings, requires also a sound knowledge of methods of making bend allowances.

If the fundamental principles are understood it is possible to simplify much of the calculation work for the majority of jobs by using a standard method of approach to bend allowance problems. For instance, by far the greater number of detail fittings are comprised of 90° bends, with usually an inside bend radius of twice the thickness of material, which makes it possible to use an easy method that has recently been evolved for the rapid calculation of such types of work.

CALCULATING POSITION OF BEND LINES

The principle of the method for 90° bends consists in calculating the position of each bend line on the pattern by deducting half the bend allowance from three times the thickness of material, and either measuring this distance back from the net pattern profile lines or subtracting it from the mould line measurement. The mould line is the intersection point, marked X in Fig. 75, of extended lines from the outside of the flanges or legs of the job.

For the simple bent work, such as is shown in Fig. 75, the pattern can be calculated directly from the dimensions given in the end view. The material used is 2 mm thick. In the pattern the bend lines are shown at B and C, and the position of each line may be calculated in the following manner.

First find length AB in the pattern by subtracting from the mould line measurement (25 mm) the difference in length between half the bend allowance ($\frac{1}{2}$ B.A.) and three times the material thickness (3G).

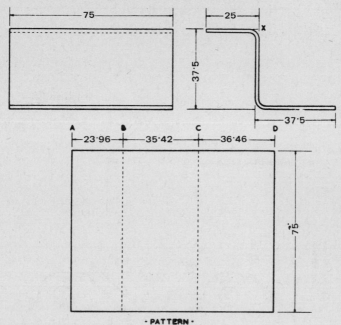

All dimensions in millimetres

FIG. 75.

Length AB = 25 − (3G − 0·5 B.A.).
   = 25 − (3 × 1 − ·5 × mean radius × 90 × ·01745)
   = 25 − (3 − ·5 × 2·5 × 90 × ·01745)
   = 23·96 mm.

Length BC is found in a similar manner, but in this case the two bends have to be taken into consideration.

Length BC = 37·5 − 2(3G − ·5 B.A.)
   = 37·5 − 2(3 × 1 − 1·96)
   = 35·42 mm.

Finally, length CD is obtained in the same way as AB.

Length CD = 37·5 − (3G − ·5 B.A.)
   = 37·5 − (3 − 1·96)
   = 36·46 mm.

All dimensions in millimetres

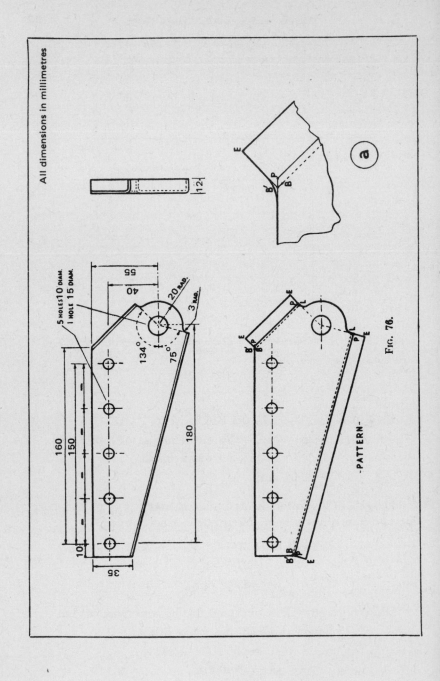

5 HOLES 10 DIAM.
1 HOLE 15 DIAM.

20 RAD.
3 RAD.
134°
75°

55
40
160
150
10
35
180
12

B'
B
P
E

PATTERN-

Fig. 76.

(a)

Later on in this Chapter 21 will be found a bend line table, in which the distance between the profile and bend lines for each gauge of material is shown. Thus, in calculating any 90° bend it is only necessary to deduct the given constant in the table for the appropriate material from the given outside dimension of the job. For instance, if the material used in Fig. 75 is assumed to be 1·2 mm thick, length AB in pattern would be obtained thus :—

$$\text{Length } AB = 25 - 1\cdot25 \text{ (from table)}$$
$$= 23\cdot75 \text{ mm.}$$

In this way bend lines can be easily calculated and marked off on to the template by means of a vernier height gauge. When inside dimensions are given on the working drawing it is necessary to add the thickness of material to the sizes in order to obtain outside dimensions.

DEVELOPMENT OF STIFFENER PLATE

Fig. 76 shows the elevation and end view of a stiffener plate made from material 1·2 mm thick, together with the construction required to mark-off the pattern. The elevation profile is drawn in a convenient position for the pattern, and allowances added for the flanges. Position of the bend lines is obtained from the bend line table. For material 1·2 mm thick, this distance from the profile lines is 1·25 mm and is marked back accordingly on the pattern.

Length BE, or LE, is calculated as follows :—

$$BE = 12 - 1\cdot25$$
$$= 10\cdot75 \text{ mm.}$$

In Fig. 76 at (a) the enlarged view of the flange edge lines illustrates the method of completing the pattern profile at this point. Line B'E is drawn at right angles to the bend line BL from point B', which is the intersection point of the flange and horizontal profile lines. To eliminate the sharp corner at point B' a slight adjustment is made to the profile by means of a free-hand curve drawn in the manner shown.

DEVELOPMENT OF TUBE BRACKET

The foregoing simplification of the marking-out process for 90° bend allowances can be successfully applied to the

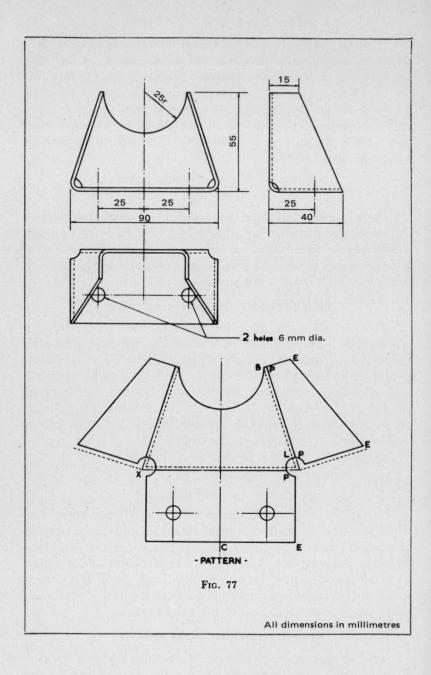

25r

55

25   25

90

15

25

40

2 holes 6 mm dia.

B P   E

E

L P

X   P

C   E

- PATTERN -

Fig. 77

All dimensions in millimetres

development of many detail fitting patterns. Fig. 77 depicts the plan, elevation, and end view of a tube bracket made from 1·6 millimetre material, together with its developed pattern. The base and side edges of the bracket are welded together. Tackle the development in the following manner.

First draw the elevation profile in a suitable position. Next obtain from the table the distance between the profile and bend lines. This distance for 1·6 mm material is 1·66 mm.

Measure back 1·66 mm from the profile lines as depicted by the dotted lines. Next, from each side of the pattern vertical centre line measure off length CE (43·4 mm), which is equal to half the base length (45 mm) minus the metal thickness. Draw lines through the points parallel to the vertical centre line. Cut them off in length from the base bend line with length EL, which is 38·34 mm long, found as follows :—

$$\text{Length EL} = 40 - 1·66$$
$$= 38·34 \text{ mm.}$$

Next find length BE and draw at right angles to the profile lines PP.

$$\text{Length BE} = 15 - 1·66$$
$$= 13·34 \text{ mm.}$$

Draw lines of indefinite length parallel to BE, shown dotted, from the intersection point of the profile lines, and measure back parallel lines a distance of one gauge. Cut these lines off in length distance EL (38·34 mm) from the bend lines, and complete the rest of the pattern as shown.

BEND LINE TABLE

For 90° bends with an inside radius of twice the metal thickness, deduct the constant for the appropriate gauge from the given *outside* measurement of the job.

Material	Nearest Metric Thickness	Bend Line Constant
10 S.W.G.	3 mm	3·11 mm
12 S.W.G.	2·5 mm	2·59 mm
14 S.W.G.	2 mm	2·08 mm
16 S.W.G.	1·6 mm	1·66 mm

18 S.W.G.	1·2 mm	1·25 mm
20 S.W.G.	0·9 mm	0·94 mm
22 S.W.G.	0·7 mm	0·73 mm
24 S.W.G.	0·6 mm	0·63 mm

DEVELOPING BEND ANGLES OTHER THAN 90°

Various types of fittings with bend angles other than 90° can be developed by calculating their stretched-out length in the flat by using a similar method to that illustrated in Fig. 70 (*b*). The bracket shown in Fig. 78 is made from material 2 mm thick, with inside bend radii of 2 gauge. At first sight it may appear simple enough to develop, but this can only be done accurately from the information given in the plan and elevation by means of the method shown at (*a*) and (*b*) in the diagram. First of all it is necessary to calculate the angle at which the centre portion of the bracket is inclined to the horizontal centre line in plan. In Fig. 78 at (*b*), triangle HDX is formed of a base 10 mm long—which is the difference in length between the side dimensions of the bracket as at (*a*)—and a perpendicular 42 mm in length. This length is found by adding 4 mm (two thicknesses of metal) to the inside width (38 mm) of the bracket.

Angle HDX can thus be found as follows :—

$$\text{tan angle HDX} = \frac{\text{HX}}{\text{DX}}$$

$$= \frac{10}{42}$$

$$= ·2381$$

hence, angle HDX $= 13° 23'$.

The exterior angle, *which is similar to the bend angle* BLE, shown at Fig. 78 (*a*), is thus 13° 23′ plus 90°, *i.e.*, 103° 23′.

Next calculate the bend allowance for angle BLE as follows :—

$$\text{Bend allowance BE} = 2·5\text{G} \times 103° 23' \times ·01745$$

$$= 5 \times 103·38 \times ·01745$$

$$= 9·02 \text{ mm.}$$

The bend allowance for angle FKI is next found by first calculating angle DHX—subtracting angle HDX (13° 23′) from 90°. Thus bend angle FKI is 76° 37′, from which the bend allowance can be found as follows :—

$$\text{Bend allowance } FI = 2 \cdot 5G \times 76° \ 37' \times \cdot 01745$$
$$= 5 \times 76 \cdot 62 \times \cdot 01745$$
$$= 6 \cdot 69 \text{ mm.}$$

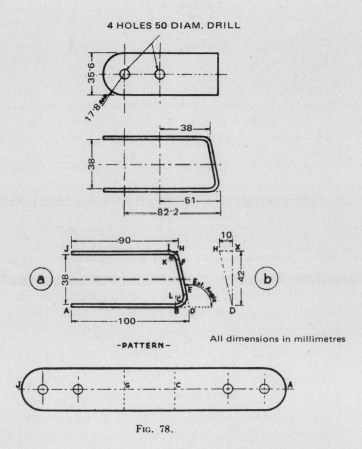

Fig. 78.

To find the length of flat AB it is necessary to calculate the length of BD. This is the base length of triangle BLD, of which the perpendicular is 3 gauge in length. Angle BLD is half-angle BLE, *i.e.*, 51° 42'.

Thus,   length BD $= \tan 51° 42' \times 3G$

$= 1·2662 \times 6$

$= 7·60$ mm.

therefore,   flat AB $=$ AD $-$ BD

$= 100 - 7·6$

$= 92·4$ mm.

Use the same method to find the length of flat IJ. First calculate HI by dividing angle FKI (76° 37') by two, to obtain 38° 19'. This is angle HKI. Perpendicular IK is 3 gauge in length.

Therefore,   length HI $= \tan 38° 19' \times 3G$

$= ·7902 \times 6$

$= 4·74$ mm.

thus,   flat IJ $=$ HJ $-$ HI

$= 90 - 4·74$

$= 85·26$ mm.

For the length of flat EF first obtain the length of hypotenuse DH as at (*b*) as follows :—

$$\text{Length DH} = \sqrt{XD^2 + HX^2}$$
$$= \sqrt{42^2 + 10^2}$$
$$= 43·17 \text{ mm.}$$

From the hypotenuse length deduct distances ED and FH, which are equal in length to BD and HI respectively.

Thus,   flat EF $=$ HD $-$ (ED $+$ FH)

$= 43·17 - (7·6 + 4·74)$

$= 30·83$ mm.

DEVELOPMENT OF PATTERN

In the pattern, distance AC is equal to the length of flat AB added to length BC, which is half of bend allowance BE.

$$\text{Length AC} = \text{AB} + \text{BC}$$
$$= 92·4 + 4·51$$
$$= 96·91 \text{ mm.}$$

Length CG is the sum of the lengths of flat EF, EC (similar to BC), and FG, which is half of the bend allowance FI.

$$\text{Length CG} = \text{EF} + \text{EC} + \text{FG}$$
$$= 30\cdot83 + 4\cdot51 + 3\cdot35$$
$$= 38\cdot69 \text{ mm.}$$

Finally, the length of GJ is the length of flat IJ plus GI, which is half of the bend allowance FI.

$$\text{Length GJ} = \text{IJ} + \text{GI}$$
$$= 85\cdot26 + 3\cdot35$$
$$= 88\cdot61 \text{ mm.}$$

The pattern is marked off in the manner shown.

DIMENSIONS TO BEND CENTRES

Care should always be taken when calculating bend allowances that working drawing dimensions are correctly identified between the appropriate extension or mould lines. For instance, in the example in Fig. 78 the dimensions are mould line dimensions, the mould line being the profile extension points shown at D and H respectively. Often, however, a working drawing shows the dimensions given to bend centres and not to mould lines.

An enlarged view of a bracket made from material 2 mm thick, dimensioned in this way, is shown in Fig. 79. In

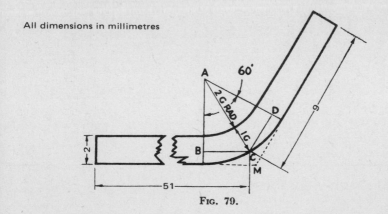

All dimensions in millimetres

FIG. 79.

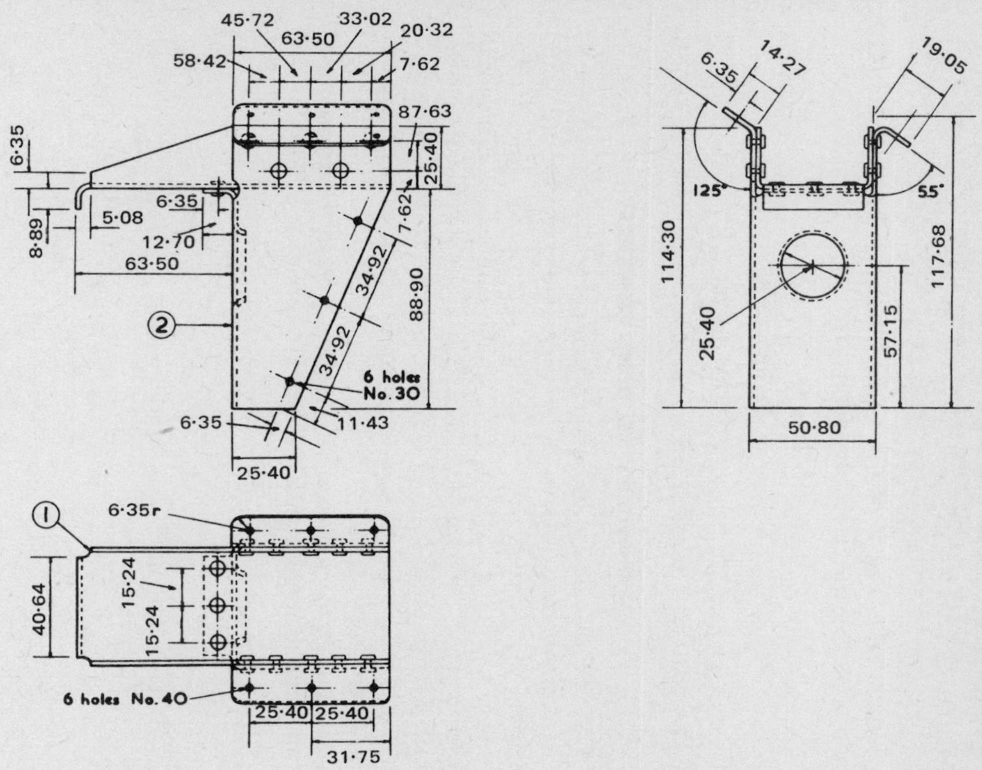

All dimensions in millimetres

FIG. 80.

this case it is necessary to calculate the length of BC in the triangle BAC in order to find the correct lengths of the flats. As the bend angle is given as 60° angle BAC is thus 30°, and as the length of AC is 3 gauge, the length of BC can be found as follows :—

$$\text{Length } BC = \sin 30° \times AC$$
$$= \cdot 500 \times 6$$
$$= 3 \cdot 0 \text{ mm.}$$

This dimension is then subtracted from 51 mm and 9 mm to obtain 48 mm and 6 mm as the length of each flat.

DEVELOPMENT OF SPAR SUPPORT BRACKET

Many detail fitting drawings appear very complex, and problems of template making can assume rather difficult proportions, yet often on analysis the majority of the developments can be accurately obtained on the simplified lines explained in this work. An example is shown in Fig. 80, which is a typical aircraft detail fitting as set-out on a working drawing. Essentially the job consists of two fittings riveted together to form a spar support. In the templates of each part all holes are shown in the flat. The developments of the patterns are depicted in Figs. 81 and 82.

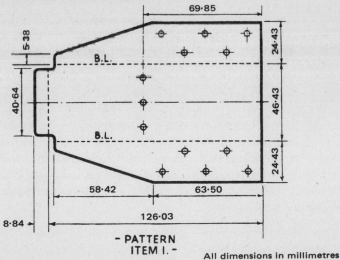

– PATTERN
ITEM I. –

All dimensions in millimetres

FIG. 81.

All the calculated sizes are shown on the patterns, and the methods of obtaining these dimensions should present no difficulty. Note carefully that the dimension of 117·68 mm given on the working drawing between the base of item 2 and the acute-angled flange is to the *mould line*, while the measurement of 114·30 mm to the other flange is to the bend centre. In the pattern for item 2 the flange bend lines are calculated to the centre of the appropriate bend allowances.

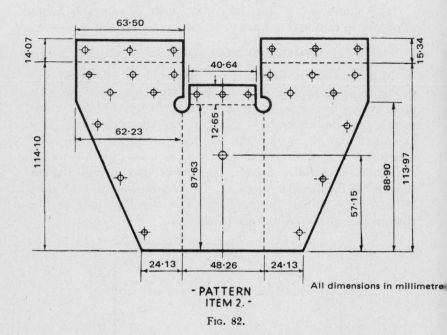

All dimensions in millimetre

- PATTERN
ITEM 2. -

FIG. 82.

Aircraft detail fitting design covers a very wide field, and new development problems, as in every other branch of sheet metal engineering, are constantly arising. By using the principles of pattern lay-out as set forth in this work a modern-minded sheet metal worker should be able to tackle any such problems with ease.

Chapter Twenty-two

IRREGULAR SURFACE PERIMETER CALCULATIONS

FINDING the correct length of material to fit round the perimeter of an irregular surface is an everyday problem in sheet metal working. Many articles, such as fuel tanks, machine guards, and covers, are of irregular shape, and the following examples illustrate how their perimeters can be calculated, taking into account allowances for the thickness of material used.

In Fig. 83 is shown an aircraft auxiliary fuel tank of welded construction. The tank is oval in shape, and is made from material 0·7 mm thick.

Referring to Fig. 83 (a), the developed length of material to form the tank shell is calculated in the following manner.

First find the angle BAC :—

$$\text{sin angle BAC} = \frac{BC}{AC}$$

$$= \frac{225}{462}$$

$$= \cdot 4870$$

hence,   angle BAC $= 29° \, 08'$.

Next calculate the length of arc E–F, including allowance for metal thickness (0·5G).

$$\text{Arc E–F} = (\text{EA} + 0 \cdot 5\text{G}) \times 29° \, 08' \times \cdot 01745$$

$$= 614 \cdot 35 \times 29 \cdot 13 \times \cdot 01745$$

$$= 312 \cdot 28 \text{ mm.}$$

Now find angle DBE :—

$$\text{Angle DBE} = 90° - 29° \, 08'$$

$$= 60° \, 52'.$$

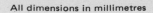

All dimensions in millimetres

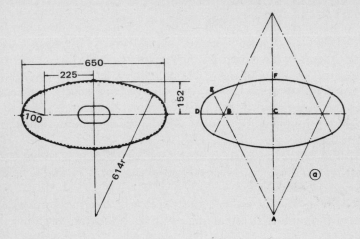

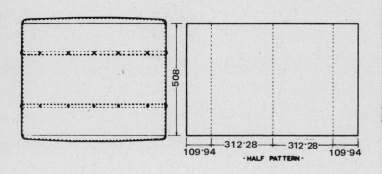

- HALF PATTERN -

Fig. 83.

Next calculate the length of arc D–E, including allowance for metal thickness.

$$\text{Arc D–E} = (\text{BD} + 0.5\text{G}) \times 60° \ 52' \times .01745$$
$$= 100.35 \times 60.866 \times .01745$$
$$= 109.94 \text{ mm.}$$

Therefore, the total length of material required to develop the half pattern is $(312.8 + 109.94) \times 2 = 845.48$ mm.

The next example, shown in Fig. 84, is of a belt guard of welded construction, made from material 1.4 mm thick. Its pattern is developed as follows :—

Referring to Fig. 84 (a), first find angle BAC :—

$$\sin \text{angle BAC} = \frac{\text{BC}}{\text{AC}}$$
$$= \frac{200}{450}$$
$$= .444$$

hence, angle BAC = 26° 23′.

Note :— BC = CE − AF = 350 − 150 = 200 mm).

Next find angle DCE :—

$$\text{Angle DCE} = 90° + 26° \ 23'$$
$$= 116° \ 23'.$$

Now calculate the length of arc D–E, including allowance for metal thickness.

$$\text{Arc D–E} = (\text{CE} + 0.5\text{G}) \times 116° \ 23' \times .01745$$
$$= 350.7 \times 116.383 \times .01745$$
$$= 712.22 \text{ mm.}$$

Find angle GAF :—

$$\text{Angle GAF} = 90° - 26° \ 23'$$
$$= 63° \ 37'.$$

Next calculate the length of arc F–G, including allowance for metal thickness.

$$\text{Arc F–G} = (\text{AF} + 0.5\text{G}) \times 63° \ 37' \times .01745$$
$$= 150.7 \times 63.617 \times .01745$$
$$= 167.29 \text{ mm.}$$

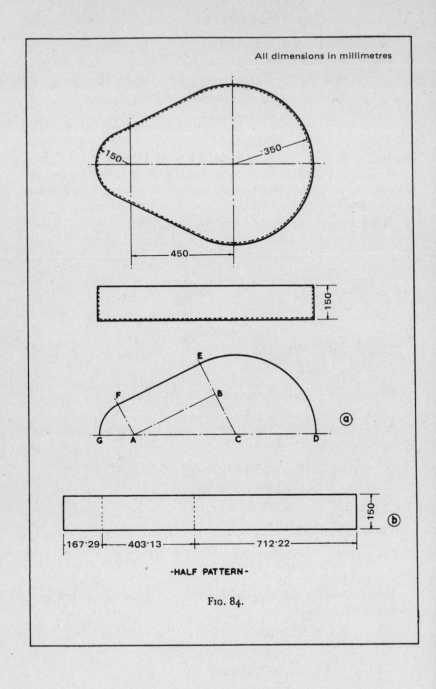

All dimensions in millimetres

-HALF PATTERN-

Fig. 84.

Finally, find the length of flat EF, which is equal to AB :—

$$\text{Flat } EF = AC \times \cos \text{ angle } BAC$$
$$= 450 \times \cdot 89584$$
$$= 403 \cdot 13 \text{ mm.}$$

The developed half pattern is then marked off as shown at (b), Fig. 84.

Another type of guard in general use is that for covering gear wheels, as shown in Fig. 85. Of welded construction, this gear guard is made of material 2 mm thick, and its pattern is calculated as follows :—

Referring to Fig. 85 (a), first find angle BAC :—

$$\sin \text{ angle } BAC = \frac{BC}{AB}$$
$$= \frac{425}{625}$$
$$= \cdot 6800$$

hence,   angle $BAC = 42° \, 51'$.

Next find angle DAF :—

$$\text{Angle } DAF = 180° - 42° \, 51'$$
$$= 137° \, 9'.$$

Next calculate the length of arc D–F, including allowance for metal thickness.

$$\text{Arc } D\text{–}F = (AD + 0 \cdot 5G) \times 137° \, 9' \times \cdot 01745$$
$$= 426 \times 137 \cdot 15 \times \cdot 01745$$
$$= 1019 \cdot 52 \text{ mm.}$$

Now find angle DBE :—

$$\text{Angle } DBE = 90° - 42° \, 51'$$
$$= 47° \, 9'.$$

Next calculate the length of arc D–E, allowing for metal thickness.

$$\text{Arc } D\text{–}E = (EB + 0 \cdot 5G) \times 47° \, 9' \times \cdot 01745$$
$$= 199 \times 47 \cdot 15 \times \cdot 01745$$
$$= 163 \cdot 73 \text{ mm.}$$

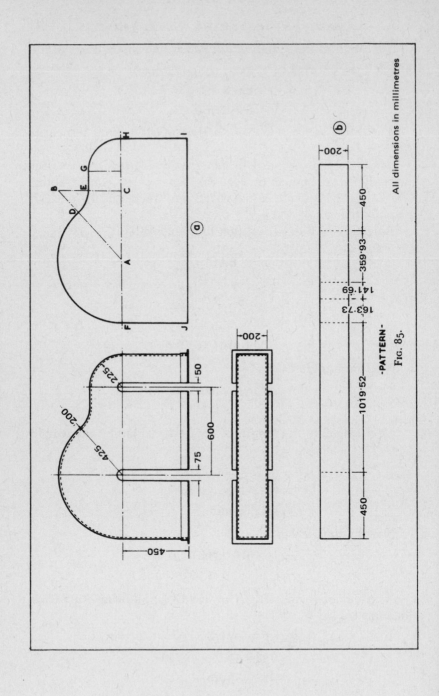

All dimensions in millimetres

-PATTERN-

Fig. 85.

To calculate the length of flat EG, it is first necessary to find the length of AC :—

$$\text{Length AC} = \text{AB} \times \cos \text{angle BAC}$$
$$= 625 \times \cdot 73314$$
$$= 458 \cdot 21 \text{ mm.}$$

thus, length of flat EG :—

$$\text{Flat EG} = 600 - 458 \cdot 21$$
$$= 141 \cdot 79 \text{ mm.}$$

Finally, calculate the length of arc G–H, including allowance for metal thickness.

$$\text{Arc G–H} = (225 + 0 \cdot 5\text{G}) \times 90° \times \cdot 01745$$
$$= 226 \times 90 \times \cdot 01745$$
$$= 359 \cdot 93 \text{ mm.}$$

The full developed length of the pattern is marked off as shown in Fig. 85 (*b*).

FINDING THE CORRECT PERIMETER LENGTHS OF PIECES WITH IRREGULAR SURFACES

Working drawings for articles of irregular surface do not usually give all the dimensions necessary for finding the correct lengths of their perimeters. An example is the cover shown in Fig. 86, which is of welded construction, made of material 1·2 mm thick. The pattern is calculated in the following manner :—
Referring to Fig. 86 (*a*), first find length AB :—

$$\text{Length AB} = \text{JK} - (\text{JH} + \text{GK})$$
$$= 43 \cdot 08 - (6 + 12 \cdot 5)$$
$$= 24 \cdot 58 \text{ mm.}$$

Next find length BC :—

$$\text{Length BC} = \text{FG} - (\text{FC} + \text{BG})$$
$$= 25 - (6 + 6 + 1 \cdot 2)$$
$$= 11 \cdot 80 \text{ mm.}$$

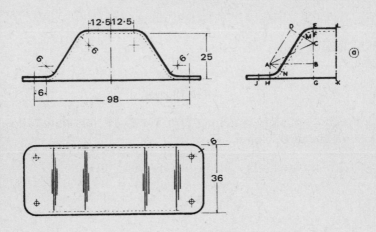

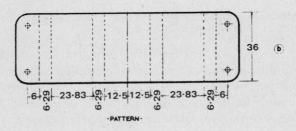

-PATTERN-

F‌IG. 86.

Then find angle BAC :—

$$\text{tan angle BAC} = \frac{BC}{AB}$$

$$= \frac{11 \cdot 8}{24 \cdot 58}$$

$$= \cdot 4800$$

hence,   angle BAC = 25° 38′.

Now find length AC :—

$$\text{Length AC} = \frac{BC}{\text{sin angle BAC}}$$

$$= \frac{11 \cdot 8}{\cdot 4326}$$

$$= 27 \cdot 24 \text{ mm.}$$

Next find angle DAC :—

$$\text{sin angle DAC} = \frac{CD}{AC}$$

$$= \frac{6 + 6 + 1 \cdot 2}{27 \cdot 24}$$

$$= \cdot 4846$$

hence,   angle DAC = 28° 59′.

Bend angle HAN, therefore, is :—

$$\text{Angle BAC} + \text{angle DAC}$$

$$= 25° 38′ + 28° 59′$$

$$= 54° 37′.$$

Next calculate the bend allowance, allowing for thickness of metal.

$$\text{B.A.} = (\cdot 25 + 0 \cdot 5G) \times 54° 37′ \times \cdot 01745$$

$$= 6 \cdot 6 \times 54 \cdot 616 \times \cdot 01745$$

$$= 6 \cdot 29 \text{ mm.}$$

Finally, find the length of flat MN, which is equal to AD :—

$$\text{Flat MN} = \text{AC} \times \cos \text{ angle DAC}$$
$$= 27 \cdot 24 \times \cdot 8748$$
$$= 23 \cdot 83 \text{ mm.}$$

The pattern is then marked off as shown in Fig. 86 (*b*).

In this example, dimensions are to the centres of the bend radii, but draughtsmen more often dimension to mould lines. Mould line dimensioning, however, directed to outside profiles, can give rise to further problems of development, as in the example of a bracket made of material 2 mm thick, shown in Fig. 87.

This development is calculated as follows :—

First find the length of AC :—

$$\text{Length AC} = \sqrt{\text{AB}^2 + \text{BC}^2}$$
$$= \sqrt{25 \cdot 5^2 + 31 \cdot 9^2}$$
$$= 40 \cdot 83 \text{ mm.}$$

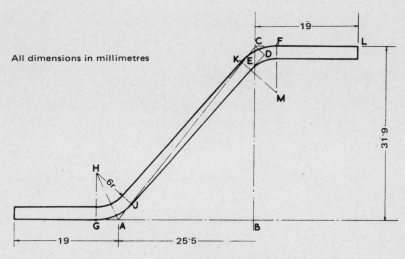

All dimensions in millimetres

Fig. 87.

Next find angle BAC :—

$$\sin \text{angle BAC} = \frac{BC}{AC}$$

$$= \frac{31 \cdot 9}{40 \cdot 83}$$

$$= \cdot 7813$$

hence,   angle BAC = 51° 23′.

Now find angle DAC :—

$$\sin \text{angle DAC} = \frac{CD}{AC}$$

$$= \frac{1 \cdot 9}{40 \cdot 83}$$

$$= \cdot 0465$$

hence,   angle DAC = 2° 40′.

Find angle EAB :—

Angle EAB = angle BAC − angle DAC

= 51° 23′ − 2° 40′

= 48° 43′.

The bend allowances and lengths of flats are then calculated, as previously explained. Note, however, that the length of AD has to be found before the lengths of flats can be calculated. Angle GHJ is the angle required for the bend allowance calculations.

If the above principles of calculating perimeters of irregular surfaces are memorised, they can be applied to the development of innumerable patterns for precision sheet metal work.

Chapter Twenty-three

SNAP JOINTS

NEWmethods of fastening the edges of sheet metal by joints which clip together by hand pressure have been introduced into the industry in recent years. Such joints have found their main uses for the side lap fastening of roofing and siding sheets and for the assembly of air duct components.

By using snap joints many sheet metal articles of a bulky nature can be easily and rapidly assembled at the point of use, without tools, which also means that savings are made in transport and storage costs. These new joints can be made in folders and press brakes, but the most efficient production is by means of cold roll forming in which the edges of the sheet are progressively formed to the correct shape and size by means of a series of rolls in a roll or lock forming machine.

Snap joints are preferably made from fairly hard light gauge material as it is essential that their construction should provide sufficient springiness to avoid the parts being overstrained as they are clipped together.

In Fig. 88 is shown a structural type of corner snap joint. It is made of 0·7 mm thick commercial quality aluminium sheet by forming its mating components, the clip and the lock, into wedge-shaped sections with inturned lips, on the edges of separate pieces of material. By using hand pressure progressively along the components the lock is pushed into the clip, involving temporary alteration of the wedge angles by reason of the interference between the cross-sectional width of the lock and the lip of the clip, until both lips engage tightly together. The interlocking lips prevent the components of the joint from being pulled apart in the lateral direction, but to release the snap lock they can be slid apart relative to each other in the longitudinal direction.

The following formula is used for providing the required amount of spring action in the construction of the structural snap joint:

Length of clip (LC) = metal thickness × 30
Inside bend radius = LC × ·2
Inside depth of lip = LC × ·125

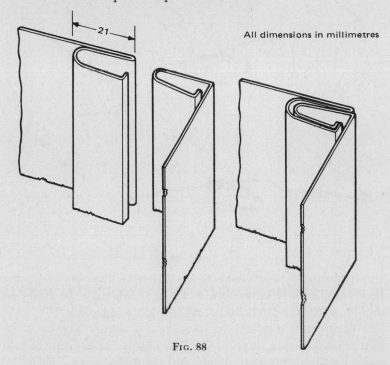

All dimensions in millimetres

Fig. 88

The outside bend radius of the lock section is made 0·5 mm
less than the inside bend radius of the clip section.

The joint is suitable for cabinet manufacture, as it provides
an internal column of considerable strength at the vertical corners of the cabinet, making unnecessary any form of framework.
Preferably made from pre-painted or plastic coated sheet metal,
the cabinets can easily be assembled by hand pressure, either in
the home or factory, without damaging the paintwork.

For round piping, as used in ventilation work, the joint
shown in Fig. 89 provides a continuous snap seam which enables
a number of open jointed pipes to be nested together for transport purposes.

Usually the piping is made of galvanised mild steel 0·6 mm
thick. The base wall of the clip section is made shorter than

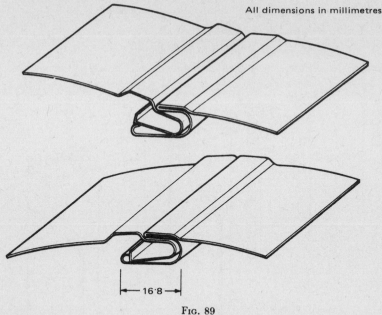

All dimensions in millimetres

|← 16·8 →|

Fɪɢ. 89

the structural type of corner joint shown in Fig. 88, as this feature enables the lock section to be partly hooked into the clip
section when the parts are pressed into engagement.

As shown in Fig. 89, the lock section rotates slightly in the
clip section during the process of assembly. This type of snap
seam joint requires very little hand pressure.

The formula for calculating the sizes of the snap seam is as
follows :

> Overall length of clip (LC) = metal thickness × 30
>
> Length of clip base   = LC × ·66
>
> Inside bend radius   = LC × ·2
>
> Length of hooked lip   = metal thickness × 6

Another type of snap joint, with a " button lock ", is shown
in Fig. 90. It is generally used for snap seaming the longitudinal
edges of rectangular or square section air ducts made of
galvanised mild steel 0·5 or 0·6 mm thick. A series of slots positioned at intervals in the free wall of the clip snap over and
interlock behind wedge-shaped buttons pressed at correspond-
ing intervals in the flange of the lock. During manufacture the

All dimensions in millimetres

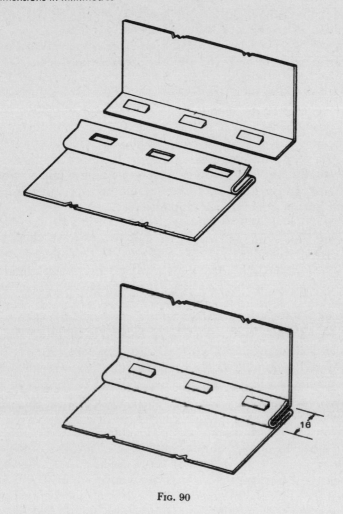

FIG. 90

raised locking edges of the buttons are cut through the flange so
that the raw edges firmly grip the edges of the slots in the clip
section when the joint is assembled.

Little pressure is needed to clip the button joint together, as
there is only intermittent frictional contact between the edges of
the buttons and the front edges of the slots. The lock section is

obviously not resilient, therefore the clip section can only be made efficiently in light gauge material which will not be over-strained when the lock is forced into it.

The maximum height of the buttons = metal thickness × 1·5

Rectangular or square section air ducts are usually made in two halves with the snap lock parts on opposite corners, so that the components will nest easily together for transport. Snap lock joints provided on the ends of ducts also enable two or more lengths to be joined together quickly and efficiently. End joints are usually made by riveting or spot welding a clip section on one end of a duct and pressing in a number of " buttons " by a hand button punch in the opposite end.

At (a) in Fig. 91 is shown a snap rib joint for sheet metal roofing and siding, made of 22 gauge (0·7 mm) aluminium, which has several features of importance. The joint provides a secure weathertight snap joint between the longitudinal side ribs of the roofing sheet, of which a cross-section is shown at (b). The interlocked ribs have considerable structural strength. Fig. 91 at (c) shows the joint ribs just prior to being snapped together. Fixings to the purlins or rails of a building are concealed. Roofing sheets with snap rib side laps are very easily and rapidly assembled on a roof structure. Wind suction loading on the sheets has the effect of tightening the joints. The joints are releasable for dismantling purposes, or for the removal and replacement of damaged sheets.

The clip rib has a semi-beaded locking edge which interlocks with a similar smaller co-operating bead at the base of the inner wall of the lock rib. Nails or screws secure the flange of the sheet to the purlins, their heads being covered, as shown at (c), when the clip rib is snapped over the lock rib. No holes have to be punched through the sheets, and consequently leakages cannot take place. A slight space is allowed between the assembled ribs to allow the spring action to function when snapping the ribs together, and to prevent capillary action. During thermal expansion and contraction of the sheets the ribs can slide slightly relative to each other. The snap rib joint is designed so that the roofing sheets will stack and nest closely together for economical transportation.

All dimensions in millimetres

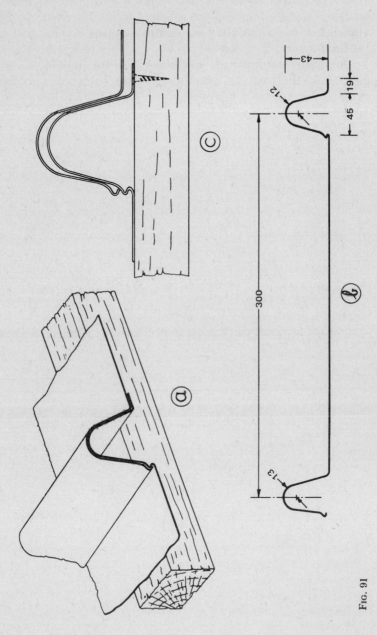

(a)

(b)

(c)

300

45 | 19

43

12

13

Fig. 91

To dismantle the sheets the rib joints can be released by inserting a wedge-shaped piece of sheet metal between the joint and sliding this along the base of the snapped ribs to open and lift them apart.

Accurate patterns are necessary for the tooling of snap joints, and these can be developed by the calculation methods explained in previous chapters.

Chapter Twenty-four

NEW DEVELOPMENTS IN COLD ROLL-FORMING TECHNIQUE

THE process of cold roll-forming sheet-metal sections of light gauge has hitherto been dominated by the concept that the moving strip must pass through a large number of roll stands, undergoing in each stand only a relatively small amount of forming before achieving its final shape in the last stand. This gradual formation of the strip through a number of carefully calculated partial cross-sections into the cross-section finally desired involves considerable roll pressures, governed only by the amount of working which the material will withstand during a given roll pass. The number and shape of the roll stands required varies widely according to the kind of material used, its thickness, and the size and intricacy of the cross-section required ; but it can be said that the technique is reasonably efficient for the production of fairly narrow profiles, the cost of the machines and rolls not being high given a reasonably large planned output.

It is only when wide sections are required—particularly if these call for considerable deep forming near the centre of the strip—that the process can become both inefficient and expensive. The large and powerful machines required for the roll-forming of wide strip call for considerable capital investment, and the rolls needed to produce the various partial profiles can cost thousands of pounds. Much expensive time is also lost between production runs by the need to change and reset the large number of rolls involved when it is desired to change the profile of the cross-section undergoing fabrication.

It has already been seen that the pressure-forming of sheet-metal strip into the desired shape in a given pair of rolls in a series of roll stands is limited by the degree of formability which the metal can stand in each roll pass without an unacceptable degree of thinning and tearing. Roll designers, absorbed in their specialist's approach to the gradual step-by-step forming of sheet metal, sometimes appear to lack the experienced sheet-metal-worker's feel for the art—often almost instinctive but derived ultimately from knowledge of the earliest handworking methods—of manipulating material into easy-flowing forms which can be, and often are, *temporarily* outside the limits of the sizes and contours of the final shape. It would seem that the cold roll-forming process itself developed from the drawbench type of forming in which a narrow strip is pulled through a series of rolls profiled for the various changes of section required to form up to the final shape. The application of power to the rolls—and so *driving* the strip through them rather than pulling it as before—probably gave rise to the roll-forming process as it is known today; but the literature on the subject is scanty, and careful search has failed to reveal any patents which have deviated from the basic concept of converting strip step-by-step into a given shape.

A SIGNIFICANT EXPERIMENT

Cold roll-forming has been described as " a process in which a flat strip of metal is progressively formed into a desired shape by passing through a series of rolls, arranged in tandem ". Although this definition is accurate enough in the sense that flat strip is used for cold roll-forming, usually as it is unwound from coil, it seems to have misled sheet-metal fabricators into taking it for granted *that the strip has to be kept flat across its entire width* before it enters the first stand of a roll-forming machine. That this assumption is not necessarily true was established in an investigation into the art and practice of cold roll-forming which took place some years ago under the author's direction in the firm which bears his name.

It was found that production of a roofing sheet of complex profile incorporating the snap rib joint shown at (*a*) in Fig. 91 was not being satisfactorily achieved from a 13-stand cold roll-forming machine. Quality of output was not always up to standard, and it was not easy to assess how much of this was caused by variations in the characteristics of the flat sheet being used, and how much to more basic shortcomings in the techniques of roll-forming as applied to thin-gauge strip of more than a certain width. It was therefore decided to attempt the development of a new type of machine which would be inexpensive to manufacture and simple to operate, yet capable of high-speed quality production of complex profiles in sheet metal of thin gauge and ordinary width.

The problem boiled down to one of finding a way of cold roll-forming a wide section from coiled strip by methods based more on the art of the sheet-metal-worker than on that of the engineer. Rolls, in short, had to be designed for a machine which could achieve the final section by allowing the metal to bend into a series of free-flowing shapes unhampered by the conventional necessity of having to undergo a progressive series of partial changes in cross-section as it passed through each roll stand.

Experiments showed that it was impossible to form a fairly deep, narrow-based rib in the middle of a wide strip of light-gauge sheet by passing it in the flat condition in which it had come off the coil through a pair of rib-forming rolls of large diameter with parallel side flanges, without thinning and splitting the material from which the rib itself had to be formed. The trouble was that the flat portions of the strip on either side of the rib, being retained in the flat condition between parallel top and bottom rollers, resisted being drawn inwards towards the rib.

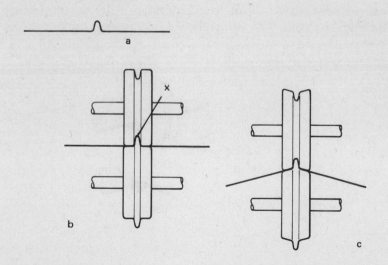

Fig. 92.

The problem, and the elements of its solution, are illustrated in Fig. 92, where the required section is shown at (*a*) and the rolls which caused the crown of the rib to split at Point X are shown at (*b*). Modified rolls were then tried, as illustrated at (*c*), where angled flanges cooperate to cause the strip to bend out of its flat condition and, as it were, to drape down on each side almost to the shafts of the bottom roll. The leading end of strip was first bent until its cross-section assumed the approximate shape of an inverted valley, and was then fed into the rolls from a low level.

Fig. 93 shows a perspective view of the essential part of the experimental set-up. Because the new procedure allowed the rolls to accept *strip already bent at its leading edge*, the full desired shape of the rib was successfully formed without any sign of unwanted thinning of the material. It was found that the temporary deformation of the strip into the transitional shapes shown in the Fig. at (*a*) and (*a*¹) provided semi-resilient skirts on either side of the rolls which allowed the material to form with considerably less restraint than had been set up when it was held flat throughout its full width.

In Fig. 93 it will be seen that the strip, bent out of the flat plane in both longitudinal and lateral directions, assumes the approximate shape of an inverted valley prior to its passage through the rolls. In the rolls, the top portion of this valley

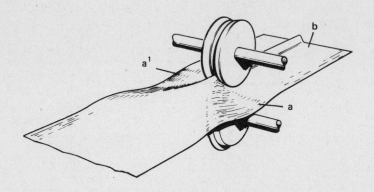

Fig. 93

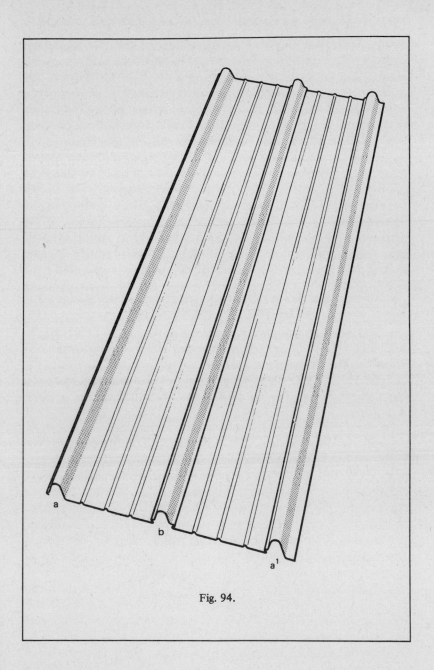

Fig. 94.

section is formed to the desired contour of the rib, while the bent portions on either side revert to the flat condition soon after passing through the rolls, as seen at (b) in the Fig.

In conventional roll-forming, the formation of the deep, narrow-based rib shown in Fig. 92 (a) would require a minimum of three roll stands, with each pair of rolls progressively deepening the rib to the desired final shape.

It would seem at first sight that the considerable amount of temporary deformation which the strip undergoes in this new method as it passes through the rolls subjects it to heavy punishment. This is not so, however, because the unconstrained strip is allowed to assume its own free-flowing shape before the position of the side guides needed to guide its entry into the first roll stand is determined. (These guides are not shown in Fig. 94 in order to clarify the principle involved.)

THE FIRST DEFORMATION-TYPE ROLL-FORMING MACHINE DESIGNED FOR PRODUCTION

The basic technique of cold roll-forming by temporarily deforming the strip in the manner described having been confirmed by the experimental set-up, an operational machine was designed to produce the sheet-metal secret-fix roofing and siding sheet illustrated in Fig. 94 and shown in detailed cross-section in Fig. 95.

Essentially, the sheet consists of two longitudinal side ribs, (a) and (a^1), comprising a snap rib-joint of the type illustrated in Fig. 91 used to secure the overlapping ribs of adjacent sheets, plus a centre rib (b) for attaching the sheet to mating clips secured to the purlins of a building. The sheet is made of aluminium alloy in coil form, either plain or with a plastic-coated finish, or alternatively from mild steel coil, either galvanised or colour-coated. The thicknesses of the coiled sheet used vary between ·5 mm and ·9 mm, and lengths up to 16 metres have been produced without trouble.

From the profile shown in Fig. 95 it was possible to calculate the exact developed width of the strip required by using the calculation methods described in earlier chapters. It turned out in this case to be 760 mm.

The roll-forming machine arrowed (a) in Fig. 96 is the basic component of a roll-forming line comprising also a flying

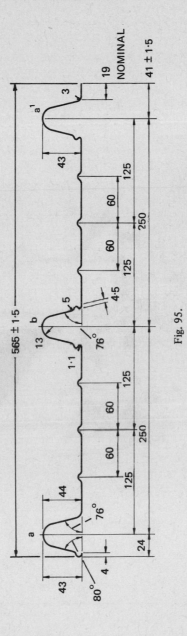

Fig. 95.

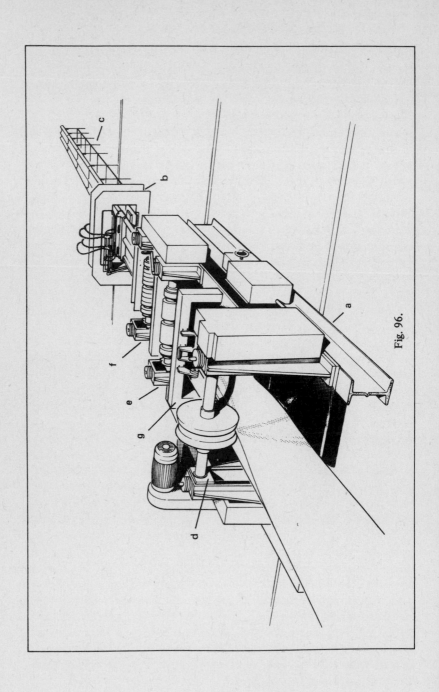

Fig. 96.

shear (b), a run-out table (c) and an uncoiler (not shown so as not to clutter up the more important parts of the illustration). The chain-driven roll-former itself consists of three roll stands, arrowed (d), (e) and (f). In the first stand (d), the deformation rolls which form the centre rib in a single pass are 460 mm in diameter, excluding their angled flanges. Stands (f) and (g) have 180 mm-diameter rolls for shaping up the outside ribs : being on the edges of the strip, these are fairly easy to form. Stand (e) is positioned 1·5 m from stand (d). In this short distance the strip, now complete with its centre rib, re-forms itself smoothly from the temporary shape into which it was forced in the deformation rolls, and equally smoothly resolves into smaller transitional inverted-valley edge shapes before entering the second set of rolls in stand (e). These edge shapes are supported by suitable formers positioned at the rolls which form the side ribs.

The bridge (g) positioned between the first and second stands serves as a mount for small auxiliary guide rolls to control the moving strip as it passes between the stands.

The rolls are designed to form the changes in section shown in sequence at (a), (b) and (c) in Fig. 97. The forming of the re-entrant–angled locking nibs for the snap action of the profile is done by vertical rolls of suitable dimensions positioned between the second and third roll stands. They also are not illustrated lest the clarity of the essential parts of the illustration be impaired.

It soon became clear in practice that a number of unwanted side-effects of the traditional method of forming the strip while held flat between parallel top and bottom rolls were obviated by the new process. One authority, for instance, has stated that " because the strip elongates slightly due to the pressure from the rolls, the pitch diameter of each succeeding pair of rolls is slightly increased so as to maintain a uniform roll pressure on the strip and prevent buckling ". In deformation cold roll-forming, however, no pressures other than those needed to drive the strip through the rolls are applied, with the result that no distortion or work-hardening of the material occurs and the strip travels at a constant speed through each pair of rolls. These are all chain-driven, and suitably geared for their purpose.

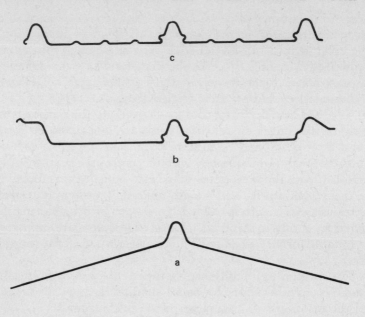

Fig. 97.

Some minor additions to the principal features of the experimental machine were found desirable. The first was to position an inner mandrel, shaped to the contour of the centre rib, between the first and second roll stands in order to counteract a tendency for the rib to collapse slightly inwardly under a pressure which was found to extend towards the rib from the edges of the formed strip.

Small vertical idler rolls were also added to control the moving strip at its edges as it passed through the roll stands; and small straightening rolls were positioned on the machine after the last roll stand in order to counteract any tendency towards spring-back in the ribs. These straightening rolls also help to overcome any tendency for side bow to develop in the coil. The sheet must be made perfectly straight for lengths of up to 16 metres. If it is not, the overlapping ribs of two adjacent sheets cannot be clipped together.

Little power is required to actuate the rolls, and a variable-speed motor of 7·5 horsepower is sufficient to produce rolling speeds of up to 36 m per minute. The flying shear shown at (*b*) in Fig. 96 needed to be carefully designed to cut the sheet off accurately to the desired lengths as it left the last roll-former stand.

The making of running adjustments to the three roll stands presents few problems and indeed is seldom necessary. The first roll stand requires no adjustment at any time.

The small number of roll stands required in the deformation roll-forming process also greatly reduces the time required for changing the rolls to form a different profile. A further saving of time can be achieved by mounting each roll stand, complete with its appropriate rolls, on a removable raft, replacing it by a similar raft with rolls already in place on its stands whenever it is desired to produce a different profile.

Lubrication by light drips of air-drying oil on the edges of the strip prevents burring on the side guides, but no surface lubricant is necessary. None of the excess metal which is sometimes contained in small buckles or ripples in the strip itself appears in the finished sheet, for the temporary deformations of the metal act to relieve tensions throughout its full width.

After a short running-in period, the first operational machine proved so successful that, with a few modifications, it was turned into a full production machine which has now been in operation for several years, turning out large tonnages of roofing sheet to a consistently high standard of precision and quality.

A MACHINE FOR ROLL-FORMING DECKING SHEET

After the success of the prototype, a second machine was designed to roll-form the light-gauge secret-fix sheet metal sheet shown in cross-section in Fig. 98. In this complex profile the two longitudinal side ribs (*a*) and (a^1) each comprise a snap rib joint. Midway between them and the centre rib *b*, two immediate-locking ribs in the surface of the sheet serve to secure it by clamp-fixing devices to the roof purlins of a building. The materials used are ·8 mm-thick aluminium alloy or mild steel, either plastic colour-coated or with galvanised finish.

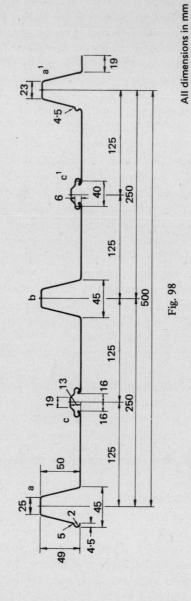

Fig. 98

All dimensions in mm

The coil of strip from which the section is formed is 785 mm wide ; and in Fig. 99 there is shown the sequence in which the required changes of section to achieve the finished profile are made in the flat strip as it comes off the uncoiler.   Four pairs of rolls are used, the first of them forming the section shown at (*a*) by the same method of allowing free-flowing " natural " deformation of the sheet to take place before it enters the roll-stand.

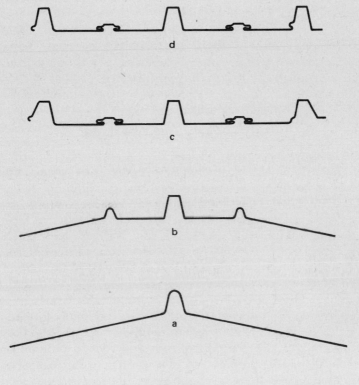

Fig. 99.

The rounded crown of the rib is next re-shaped to the de-sired trapezoidal cross-section, shown at (*b*), in the second pair of rolls, which simultaneously form two smaller ribs with rounded crowns—the material to form these ribs being drawn easily in from the draped edge regions of the strip, each shape being supported on its way to the rolls by formers positioned in front of the rolls.

Between the second and third roll stands which form sections (*b*) and (*c*), free-running vertical rolls partly form the snap-joint portions of the ribs ; and in the third roll-stand itself the small inside ribs are finished to the desired section, and the outside ribs are partially formed at the same time. These outside ribs are fully formed by other vertical rolls positioned between the third and fourth roll stands, with the last pair of rolls producing the finished profile.

The diameters of the first pair of rolls are 500 mm, and those of the other three rolls in the machine 180 mm. Suitably-positioned side gauges control the passage of the strip from the uncoiler up to the first stand, and formers (as has been seen) are provided in front of any rolls where the material needs support before entering the rolls.

SUMMARY OF ADVANTAGES

The process of cold roll-forming by temporary deformation of the strip numbers among its principal advantages that of being able to produce many sections in different gauges of material in the same rolls, with only the appropriate adjustments being necessary to the top roll settings. A specific example will illustrate why this is so. With the new methods, the bend lines at the base of a number of spaced ribs in a roofing sheet remain parallel throughout their passage through every pair of rolls and never converge between one roll stand and another. Because the strip is not being steadily drawn inwards step by step in a number of stands before it is finally formed into the desired profile of the ribs, the bend lines are as far as possible determined in the first pair of rolls and remain parallel as the material passes through all succeeding rolls regardless of whether further shaping of the preceding transitional sections takes place in these rolls or not.

Second, the comparatively small number of rolls required in the deformation roll-forming process means that scurfing and scoring of the strip cease to be hazards, and pre-painted coil can thus be roll-formed without difficulty.

Third, the completeness and comparative cheapness of deformation cold roll-forming machines can often justify the making of special-purpose machines tooled for the sole purpose of producing sections of a single shape from a fairly wide range of material thicknesses. In this way, much time is saved in tool changing and resetting, and sections can be consistently and accurately produced in very large volume.

Fourth, one of the attractions of the conventional cold roll-forming process has always been its versatility in the economic production of a variety of fairly narrow-width sections from relatively inexpensive roll-formers. The new deformation-rolling methods have been successfully applied to the manufacture of sections of this type also; and a wide range of products made from wide coil—roofing and siding sheets, flooring, shuttering, etc.—can now be made in large volume with only limited capital investment being required for the basic equipment.

Finally, reliability and durability have by now been well proven; and deformation roll-forming machines producing a wide variety of profiles have been in constant use for many years in various parts of the world.